AF812385

Hardware SecurityChallenges and Solutions

Ashutosh Mishra • Mrinal Goswami
Manoj Kumar • Navin Singh Rajput
Editors

Hardware Security: Challenges and Solutions

Springer

Editors
Ashutosh Mishra
Department of Electrical and Electronics Engineering
Birla Institute of Technology & Science
Dubai, United Arab Emirates

Manoj Kumar
University of Wollongong in Dubai
Dubai, United Arab Emirates

Mrinal Goswami
Faculty of Engineering
Assam Downtown University
Guwahati, Assam, India

Navin Singh Rajput
Indian Institute of Technology BHU
Varanasi, Uttar Pradesh, India

ISBN 978-3-031-81212-5 ISBN 978-3-031-81213-2 (eBook)
https://doi.org/10.1007/978-3-031-81213-2

© The Editor(s) (if applicable) and The Author(s), under exclusive license to Springer Nature Switzerland AG 2025

This work is subject to copyright. All rights are solely and exclusively licensed by the Publisher, whether the whole or part of the material is concerned, specifically the rights of translation, reprinting, reuse of illustrations, recitation, broadcasting, reproduction on microfilms or in any other physical way, and transmission or information storage and retrieval, electronic adaptation, computer software, or by similar or dissimilar methodology now known or hereafter developed.
The use of general descriptive names, registered names, trademarks, service marks, etc. in this publication does not imply, even in the absence of a specific statement, that such names are exempt from the relevant protective laws and regulations and therefore free for general use.
The publisher, the authors and the editors are safe to assume that the advice and information in this book are believed to be true and accurate at the date of publication. Neither the publisher nor the authors or the editors give a warranty, expressed or implied, with respect to the material contained herein or for any errors or omissions that may have been made. The publisher remains neutral with regard to jurisdictional claims in published maps and institutional affiliations.

This Springer imprint is published by the registered company Springer Nature Switzerland AG
The registered company address is: Gewerbestrasse 11, 6330 Cham, Switzerland

If disposing of this product, please recycle the paper.

Foreword

The dawn of the digital era has fundamentally transformed how we live, communicate, and interact with technology. Hardware, the bedrock of all digital systems, now faces unprecedented security challenges that could affect industries and individuals alike. As connected devices and autonomous systems continue to integrate into our daily lives, the vulnerabilities within hardware have become critical to address, requiring both innovation and rigor in security practices. This book, "Hardware Security: Challenges and Solutions," presents a timely response to these challenges. This volume explores the frontier of hardware security across a spectrum of applications, from smart vehicles to IoT ecosystems, and from cryptographic accelerators to secure supply chains. Each chapter has been crafted by experts who offer deep insights into specialized areas of security, blending theoretical frameworks with practical solutions and real-world case studies. The topics range from the role of AI in autonomous vehicle security to the revolutionary potential of blockchain for hardware trust, making this book a valuable resource for both new learners and seasoned professionals in the field. As we navigate the complexities of digital security, it becomes clear that hardware must be fortified to protect against sophisticated cyber threats. This book emphasizes that hardware security is not only a technical challenge but also a societal imperative. By advancing our understanding of the risks and solutions associated with hardware vulnerabilities, *Hardware Security: Challenges and Solutions* serves as both a guide and a call to action for continued research, collaboration, and innovation. It is a privilege to recommend this book to readers seeking to deepen their knowledge of hardware security. May it inspire further breakthroughs and contribute meaningfully to the safeguarding of our interconnected world.

Ashutosh Mishra

Preface

As technological advancements continue to drive the rapid expansion of interconnected devices, autonomous systems, and digital communication platforms, the importance of hardware security has become paramount. The growing reliance on Internet of Things (IoT) devices, smart vehicles, healthcare monitoring systems, and integrated circuits underscores the need for robust mechanisms to protect sensitive data, ensure system integrity, and prevent unauthorized access. In this context, *Hardware Security: Challenges and Solutions* brings together a wealth of research and practical insights into safeguarding hardware systems across a range of contemporary applications.

This volume explores the fundamental principles, emerging technologies, and real-world implementations that are shaping the landscape of hardware security today. Chapter 1 provides an in-depth look at AI-enabled smart vehicles, detailing the essential components, architectures, and ethical challenges involved in protecting automotive hardware. The transformative potential of blockchain technology for secure, transparent hardware systems is discussed in Chap. 2, emphasizing its role in reinforcing trust within digital ecosystems. The hardware security needs of IoT applications, where vulnerability to cyber threats poses significant risks, are addressed in Chap. 3. It presents a holistic view of security challenges in IoT and the effectiveness of hardware-assisted solutions. The security of the supply chain and manufacturing, particularly for System on Chip (SoC) and IoT edge nodes, is discussed in Chap. 4, providing insights into centralized security IP as a promising solution. In Chap. 5, attention turns to cryptographic hardware, focusing on protecting cryptographic keys from power analysis attacks through threshold-based implementations for IoT devices.

Subsequent chapters tackle diverse yet interconnected domains of hardware security. Chapter 6 highlights honeypots in intrusion detection systems, offering practical solutions to track and mitigate cyber threats. The potential of hardware-assisted blockchain for healthcare IoT applications is examined in Chap. 7, revealing how it can secure sensitive patient data and improve efficiency. Chapter 8 introduces Wireless Physical Layer Security (WPLS) as a supplementary approach to traditional cryptography, addressing the unique challenges of securing wireless communication.

The specific vulnerabilities in smart home IoT devices and the challenges of maintaining their security are discussed in Chap. 9, which bridges theoretical concepts and experimental applications. Chapter 10 presents an in-depth analysis of IoT security architectures, attack vectors, and policies from a global perspective, emphasizing the pressing need for a comprehensive approach to protect IoT ecosystems. Finally, Chap. 11 sheds light on the pervasive issue of hardware Trojans in integrated circuits, examining methods for their detection and prevention. Through its exploration of these topics, this book serves as an essential resource for researchers, engineers, cybersecurity professionals, and students seeking to understand and address the complexities of hardware security. With each chapter written by experts in the field, this volume not only emphasizes the importance of securing hardware but also charts a path forward, highlighting the innovative technologies and strategies that will play a crucial role in fortifying our digital future.

Dubai, UAE — Ashutosh Mishra
Guwahati, Assam, India — Mrinal Goswami
Dubai, UAE — Manoj Kumar
Varanasi, Uttar Pradesh, India — Navin Singh Rajput

Acknowledgements

The creation of this book, "Hardware Security: Challenges and Solutions," has been a collaborative journey, made possible by the support, insights, and expertise of numerous individuals and institutions. First and foremost, we express our gratitude to the contributing authors and researchers whose extensive knowledge and innovative research have been pivotal in shaping this volume. Their dedication and commitment to advancing hardware security have enabled us to present a comprehensive overview of contemporary challenges and solutions in this rapidly evolving field. We are deeply appreciative of the support provided by academic and research institutions, as well as industry partners who have contributed valuable resources, guidance, and feedback throughout the development of this book. Their encouragement and resources allowed us to delve into specific security areas and explore emerging trends in detail. We would also like to acknowledge our colleagues and mentors for their continued support, motivation, and intellectual contributions, which enriched the content and quality of this work. Special thanks to our publishers for their expertise and assistance in bringing this book to fruition, and to our families, whose unwavering patience and encouragement have been our foundation during this endeavor. This work is dedicated to the researchers, engineers, and cybersecurity professionals who work tirelessly to secure our digital world and protect critical infrastructures. It is our hope that this book serves as a valuable resource, fostering further research and innovation in hardware security.

Contents

Author Biographies

Ashutosh Mishra has received his B.Tech. in Electronics and Communication Engineering from Uttar Pradesh Technical University, Lucknow, India, in 2008. He completed his M.Tech. in Microelectronics and VLSI Design from the National Institute of Technology, Allahabad, India, in 2011, and his PhD in Electronics Engineering from the Indian Institute of Technology (BHU), Varanasi, India, in 2018. He has served as an assistant professor in Electronics and Communication Engineering at the National Institute of Technology, Raipur, India. Dr. Mishra was awarded the Korea Research Fellowship (KRF) in 2019 by the National Research Foundation of Korea through the Ministry of Science and ICT, South Korea. He also worked as a Brain Pool Fellow in the Seamless Transportation Lab at Yonsei University, South Korea. Currently, he is an assistant professor in the Department of Electrical and Electronics Engineering at BITS Pilani, Dubai, UAE. His research interests include smart sensors, intelligent systems, autonomous vehicles, convergence technology, and artificial intelligence.

Mrinal Goswami received his B.Tech. degree in Computer Science and Engineering from the North Eastern Regional Institute of Science and Technology, India, in 2012 and the M.Tech. and Ph.D. degrees in Computer Science and Engineering from the National Institute of Technology, Durgapur, West Bengal, India, in 2014 and 2019, respectively. He is now associated with Department of Computer Science and Engineering, Assam Down Town University, Guwahati, Assam, India, and is working as an associate professor. He has published more than 21 international research articles. He has research interests in design, testing, security, and fault-tolerance issues in digital systems and emerging nanotechnology (QCA), reversible logic, and quantum computing.

Manoj Kumar obtained his Ph.D. Computer Science from the Northcap University, Gurugram. He did his B.Tech in computer science from Kurukshetra University. He obtained M.Sc. (Information Security and Forensics) degree from ITB, Dublin, and M.Tech. from ITM University. Mr. Kumar has 11+ years of experience in research and academics. He published over 70 publications in reputed journals and conferences. He published two books and five patents with his team. He is also editing eight books from reputed publishers. Presently, Mr. Kumar is working on the post of Associate Professor (SoCS) at the University of Wollongong in Dubai. He is a member of various professional bodies and reviewed for many reputed journals. He is an editorial board member in the *International Arab Journal of Information Technology* (IAJIT), Jordon, and the *Journal of Computer Science Research*, Singapore. He is recognized as Quarterly Franklin Member (QFM) by London Journal Press from March 2019 onward and as a Bentham Ambassador (India) on behalf of Bentham Science Publisher. He delivered various key speeches and talks in national and international forums.

Navin Singh Rajput received an M.E. degree in communication systems in 1997 and a Ph.D. degree in the area of intelligent data analysis and pattern recognition techniques from the Indian Institute of Technology (BHU), Varanasi, in 2011. He is currently an associate professor with the Department of Electronics Engineering, Indian Institute of Technology (BHU), Varanasi, India. His research interests include intelligent gas sensors and systems, embedded and real-time artificial intelligence-based systems, cyber-physical systems, and digital twins for smart cities. He has authored books and e-content on smart systems and over 60 publications in refereed journals and conferences. He is now developing intelligent air quality monitoring systems for widespread commercialization at national and international scales as a spin-off company.

Hardware Security in AI-Enabled Smart Vehicles: Key Components and Dynamics

Nitin Kumar Chauhan, Amit Kumar, Krishna Singh, Sneha Nagar, and Aditya Shastri

1 Introduction

Smart vehicles are transforming the way we travel and providing insight into the next generation of transportation. Modern sensors, cameras, connection technologies, and robust hardware security measures are integrated into these intelligent devices, enabling them to interact securely with their peers and their surrounding system. Smart vehicles, ranging from self-driving automobiles to vehicles that deliver real-time traffic alerts and assist with parking, are making our roads safer, more effective, and more accessible [1]. The era of intelligent transportation promises to alter driving and revolutionize travel due to the rapid growth of technology.

The accelerated growth of AI technology is driving an unprecedented revolution in the automobile sector. AI-enabled smart cars are revolutionizing transportation by improving consumer experience, efficacy, and overall security [2, 3]. Vehicles are equipped with the ability to detect, analyze, and react to their surroundings with human-like intelligence thanks to AI and related subfields, including machine learning (ML), deep learning (DL), computer vision (CV), and natural language processing (NLP) [4, 5].

The major goal of this work is to offer a thorough grasp of the various hardware security aspects in basics, design concerns, modeling methodologies, and applications of AI in intelligent vehicles. The objective of this chapter is to illuminate the potential of AI to transform the automotive industry and shape the future of transportation by examining every aspect of AI technology, involving the implementation of comprehensive hardware security measures.

N. K. Chauhan · A. Kumar (✉) · S. Nagar · A. Shastri
Department of ECE, Indore Institute of Science and Technology, Indore, India

K. Singh
G.B. Pant DSEU Okhla-I Campus (Formerly GBPEC), New Delhi, India

© The Author(s), under exclusive license to Springer Nature Switzerland AG 2025

A. Mishra et al. (eds.), *Hardware Security: Challenges and Solutions*,
https://doi.org/10.1007/978-3-031-81213-2_1

The main ideas of AI, such as ML, DL, CV, and NLP, as well as how they apply to secure the hardware of intelligent vehicles, are covered in this chapter. It looks at the hardware needs, software design, communication, security, and privacy concerns while designing AI integration into vehicles. Additionally covered in this chapter are specialized modeling approaches utilized in smart vehicle perception, decision-making processes, and machine–human interaction. Additionally, it examines a variety of AI-based smart vehicle applications, including self-driving vehicles, smart transit systems, in-car customization, and vehicle-to-everything (V2X) connectivity. The chapter also discusses the difficulties and potential prospects for AI in the automobile sector.

The block diagram in Fig. 1 provides a visual representation of how artificial intelligence is integrated into the hardware security of smart vehicles, from sensors capturing data from the external environment to AI algorithms influencing the control of the vehicle and user interaction.

Sensors Sensors serve a substantial part in the hardware security of AI-powered intelligent vehicles, offering many layers of safeguarding and surveillance. They detect breaches and prevent unlawful entry by monitoring numerous parameters and controlling access with biometric data sensors. Sensors enable safe connection and secrecy, preventing data manipulation and snooping. Persistent surveillance and log keeping provide real-time monitoring, allowing security concerns to be detected and responded to immediately [6].

Sensors limit physical manipulation by sensing the illegal access to crucial hardware elements and issuing notifications. They also change security precautions based on the surrounding circumstances, such as changing protocols during inclement weather, and identifying outside hazards, implementing proactive measures as appropriate. Sensors, when linked with AI platforms, improve decision-making by offering useful information, allowing for proactive upkeep to avert hardware breakdowns, and assuring secured vehicle-to-everything (V2X) connections [7]. Additionally, sensors support user identification and security by keeping an eye on the driver's conduct and health, guaranteeing functioning safely, and protecting the passengers by making sure they are all securely belted up.

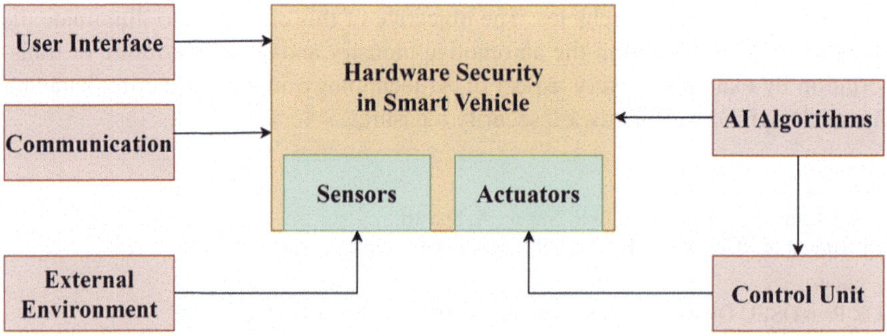

Fig. 1 AI utilization in hardware security of smart vehicles

AI Algorithms AI methods strengthen the hardware security of intelligent vehicles by utilizing improved identification, reactions, and preventive techniques. They examine sensor data to discover abnormalities and anticipate hardware breakdowns, allowing them to identify threats in real time and perform scheduled repairs. AI-powered systems identify and react to illegal access and digital hazards, providing secure interaction via encrypted and authenticated methods.

Adaptable security mechanisms enable AI to modify protocols relying on the vehicle's environment and constantly regulate privileges. By combining sensor data, AI delivers complete surveillance and detects complicated hazards. AI also protects vehicle-to-everything (V2X) connectivity by avoiding intercepted information and enabling secure connections with external equipment [8]. Additionally, AI uses biometric identification for security reasons and analyzes driving patterns for safety, thereby preserving confidential information within the car.

Control Unit The control module is critical to the hardware security of intelligent vehicles, acting as a central point for administering and implementing security regulations. It detects odd activity and illegal entry, prompting fast reactions to threats. By authenticating and validating information, a control component provides confidentiality and authenticity. It combines sensory information as well as uses AI models to detect and eliminate security hazards, allowing for improved safety actions.

The controller also administers secured via the Internet hardware and software upgrades, making sure the most recent safety updates are implemented and upgrade authenticity is verified. It also safeguards vehicle-to-everything (V2X) connectivity, guarding against malware and illicit access. It recognizes and reacts to physical manipulation efforts and does periodic self-examination to ensure hardware security. By connecting with biometric and identification networks, the control panel enables approved vehicles to be used while supporting driver surveillance systems to improve security and privacy [9].

Actuators Actuators are critical for intelligent vehicle hardware safety because they provide tangible reactions to electrical commands. They operate locking systems, freeze the car during risks, and emit alerts when illicit entry is identified. Actuators collaborate with sensors and the control module to guarantee synchronized safety reactions and adjust vehicle elements depending on real-time information [6]. They perform encrypted instructions and report their findings to the control system, ensuring ongoing surveillance. Furthermore, actuators recognize and react to physical manipulation, increase security measures, and improve user security and ease via biometric identification and autonomous driving capabilities.

Communication Communication is critical to the hardware integrity of intelligent vehicles, allowing for safe data flow among modules and exterior systems. It supports Vehicle-to-Everything (V2X) connectivity for security purposes and combines sensors and actuators with the control system to provide synchronized reactions. Communication additionally allows distant control and surveillance via telematics networks, which ensures scheduled upkeep and enhanced safety. It protects information integrity and confidentiality by encrypting and authenticating information while preventing cyberattacks. Communication also allows for secured firmware and software upgrades via the Internet (OTA), as well as urgent services and support, which improves comprehensive driving security and anonymity [8].

External Environment The external setting has an enormous effect on the hardware security of intelligent vehicles. Severe weather and unsafe driving circumstances can have an impact on sensor and actuator functionality, while Internet access and secure infrastructure might influence vulnerability to malware attacks. The ambient sensors aid in the adaptation of safety precautions to surrounding circumstances; however, interruption and dependability difficulties in V2X communication might disrupt safe data flow. Legal conformity, parking safeguards, and public consciousness are also important in safeguarding the confidentiality of smart vehicles against both external and digital hazards from the outside world [10].

User Interface The user interface (UI) in intelligent vehicles is crucial to secure hardware since it allows for user authorization via fingerprints or credentials and manages safety measures' configurations. It provides real-time surveillance of safety mechanisms and notifies users about possible attacks, as well as an instructional tool promoting optimal safety procedures. The UI connects to additional car networks, allowing for versatility and interaction with external gadgets. Furthermore, it offers accessibility to assistance and vehicle surveillance in the event of an intrusion, rendering it a critical component in guaranteeing the safety of technologically advanced vehicles [11].

Data Flow Data flow control is essential for maintaining the security of the hardware in automated vehicles. It entails safe movement and analysis of information between elements, such as encoding, authentication, and information consistency validation. Intrusion detection platforms look for irregularities in the movement of data, whereas authorization techniques guarantee that only authorized parties have access to information. Data flow policies also follow privacy rules, such as the General Data Protection Regulation (GDPR), and involve monitoring and audits for forensic inspection. Integrating safety measures allows synchronized responses to hazards, which improves security at all levels [12].

This chapter seeks to give readers a firm basis for understanding and leveraging the power of AI in influencing the next-generation transport system by conducting this comprehensive investigation of AI implementations in automated vehicles.

The chapter is structured as follows: The recent pieces of literature related to the advancement in hardware security of smart vehicles are covered in Sect. 2. The basics of AI are covered in Sect. 3, along with a summary of its history, ML methods, DL structures, CV algorithms, and NLP methods. The design concerns for integrating AI in intelligent vehicles are covered in Sect. 4 and include hardware specifications, software design, networking, and security and privacy concerns. The modeling approaches used to include AI in security aspects of smart vehicles are covered in Sect. 5, covering perceptual and contextual comprehension, decision-making and surveillance, and interaction between humans and machines. The diversified uses of AI in automated vehicles are examined in Sect. 6, including self-driving vehicles, smart transit systems, in-car customization and support, and vehicle-to-everything (V2X) connectivity. In Sect. 7, the chapter discusses ethical issues, legal frameworks, data secrecy and confidentiality, sustainability and convergence, consumer confidence and adoption, and the foreseeable possibilities of hardware safety in intelligent automobiles. It also discusses the obstacles and potential futures of AI in the automotive sector. The study finishes in Sect. 8, which highlights the revolutionary possibilities of AI in smart cars and highlights its key accomplishments.

2 Recent Advancements

Table 1 comprises a summary of recent technological developments in smart vehicles.

3 Fundamentals of Artificial Intelligence

The key components of AI are shown in Fig. 2.

3.1 AI Concept and Development

AI describes the creation of computer systems that are capable of carrying out activities that traditionally call for human intellect. These include the abilities to learn, think critically, solve problems, perceive, and comprehend language. In the middle of the twentieth century, scientists started looking into the idea of building computers that might mimic human intellect. Because of improvements in processing power, technological advancement, and the accessibility of enormous quantities of data, AI has greatly changed over time [21].

3.2 AI Concept and Development

A significant area of AI called ML concentrates on creating designs and methods that let machines acquire knowledge from information without getting specifically programmed. The trends and associations that ML algorithms discover in the data are used to inform their estimations and choices. ML comes in a variety of forms, such as supervised learning, unsupervised learning, and reinforcement learning [22].

Table 1 Recent technological advancements in smart vehicles

S. no.	Authors	Summary
1	H. Alqahtani et al. [11]	This study examines how ML may affect electric and flying vehicle safety. The research examines critical topics like intelligent decision-making, cyberattack detection, and predictive maintenance and draws important insights that will shape this technology.
2	K. Ramalakshmi et al. [12]	This article examines how AI can alter autonomous vehicles by improving travel experiences, traffic optimization, fuel consumption, and passenger safety. AI-driven autonomous vehicles in industrial, agriculture, transit, and military uses require solid cybersecurity to preserve security and confidentiality.
3	S. T. Banafshehvaragh [13]	Cyberattacks and smart automobiles are examined in this study. Investigating inter-vehicle, ground vehicle power station, in-vehicle, and IoD sensing technologies. From research 2018–2022, intrusion, anomaly, attack, and hybrid were detected and compared to smart car cybersecurity and future research solutions.
2	K. Alfrink et al. [14]	As public agencies use AI more, transparency is sought. Past studies have described AI accessibility as knowledge that helps professionals and the public make informed AI use and regulation decisions.
3	A. Karnati et al. [15]	This study likely addresses AI and self-driving cars, including purposes, impacts, and concerns. Autonomous vehicles could increase mobility for different groups, reduce traffic, and make roadways safer.
4	P. Kaur et al. [16]	Due to the rising number of cars on the road, it is harder to recognize a vehicle and its driver when they are breaching the law. ALPR is essential for security and service delivery in smart cities.
5	G. Bathla et al. [4]	Automotive Intelligent Automation (IA) uses AI and automated robotic procedures to digitally modify self-driving automobiles. IA could replace humans with automation, boosting safety and vehicle intelligence. This study compares the latest IoT, AI, and machine learning technologies for autonomous cars.
6	X. Di et al. [17]	In the mixed autonomy age, transportation engineering could use artificial intelligence (AI) models and techniques to control autonomous vehicles (AVs). Current AI-guided methodologies, problems, potential, and unsolved concerns will be discussed. AI will also help identify building blocks and locations where hybrid autonomy could benefit.
7	G. Biggi et al. [18]	This study on self-driving automobile development is scientometric and bibliometric. It evaluates the impact of AI like machine learning, deep learning, and data mining on self-driving car development.
8	Y. Ma et al. [19]	The study summarizes key works on this topic to fill the gap. Their role in supporting AVs' key AI applications is examined: (1) Perception; (2) Mapping/location; (3) Decisions. It examines current approaches to identify AI's possible uses and implementation challenges.
9	H. Thadeshwar et al. [20]	The work suggests an autonomous car that uses intelligence to operate and navigate. Self-driving capabilities are demonstrated via a 1/10-size radio-controlled car. Pi cameras, Arduino, ultrasonic sensors, CNNs, and Raspberry Pi Board type B+ are necessary.

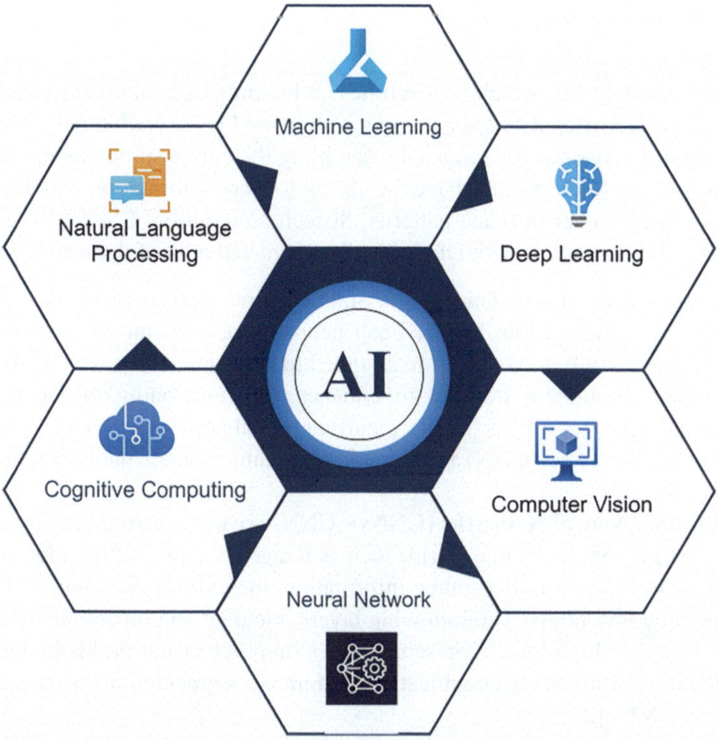

Fig. 2 Key components of artificial intelligence

Supervised Learning In supervised learning, an algorithm is trained using labeled data when its anticipated outcomes or labels are known in advance. By extrapolating from the given instances, the framework gains the ability to translate inputs into outputs. Decision trees (DT), support vector machines (SVM), and neural networks are supervised learning methods.

Unsupervised Learning With unlabeled information, unsupervised learning aims to identify fundamental frameworks or trends. This form of learning is frequently used for clustering, dimensionality reduction, and anomaly detection. Principal component analysis (PCA), autoencoders, and k-means clustering are examples of popular unsupervised learning methods.

Reinforcement Learning Through reinforcement learning, an agent may learn how to interact most effectively with its surroundings to maximize incentives. The environment responds to the agent's behavior by giving it rewards or punishments. The agent gains the ability to choose wisely to accomplish long-term objectives through experimentation and failure. The application of reinforcement learning has proved effective in fields including robotics, automated systems, and game development.

3.3 Deep Learning

DL is a subclass of ML which concentrates on building deep neural models having multiple layers. Artificial neural networks with several layers, commonly referred to as deep neural networks, are modeled resembling the composition and operation of the human brain. Such systems have the ability to learn information with hierarchical forms and uncover intricate patterns. Speech recognition, CV, NLP, and other areas have shown success with DL, which has attracted a lot of demand [22].

Neural Networks Artificial neurons or units, the linked nodes that make up neural networks, are arranged into layers. Each neuron processes inputs, activates, and then generates an output. Weights are assigned to connections between neurons and are modified throughout training to enhance the functioning of the network. Feedforward neural networks (FNN), recurrent neural networks (RNN), and convolutional neural networks (CNN) are examples of common neural network topologies.

Convolutional Neural Networks (CNNs) CNNs are made particularly for analyzing information that looks like a grid, such as images or patterns. In order to derive local characteristics from the source information, they utilize convolutional layers, which employ tiny filters. The following layers integrate and process these characteristics to create high-level representations. Computer vision problems including object identification, image classification, and image segmentation have been transformed by CNNs [23].

Recurrent Neural Networks (RNNs) Recurrent links are used in RNNs for processing data in sequence, allowing data to survive over several time intervals. RNNs can now recognize relationships and patterns in sequences as a result. Automatic speech recognition, robotic translation, and linguistic modeling are prominent applications for RNNs.

Generative Adversarial Networks (GANs) A generator and a discriminator are the two neural networks that make up a generative framework known as GANs. While the discriminator part tries to tell the difference between actual and created data, the generator segment learns to create novel instances of data that replicate the training information. GANs are useful for jobs like image production and data enhancement because they can produce realistic and excellent artificial information via an adversarial training procedure.

3.4 Computer Vision

The goal of the AI discipline of CV is to make it possible for computers to derive useful data from visual input, such as pictures and movies. It involves activities including recognizing objects, picture interpretation, monitoring, and recognition of images [24]. To analyze and comprehend visual information, visual computing algorithms use approaches from ML, pattern recognition, and computational imaging.

Image Processing Image processing methods are employed to prepare and improve images before they are put into AI models for processing. Noise diminution, image resizing, image filtration, image segmentation, and image augmentation are some of these methods.

Object Detection and Recognition Object detection and recognition techniques identify and locate objects in images or videos. They enable applications like autonomous driving, surveillance systems, and augmented reality since they can discriminate between and categorize various object types.

Semantic Segmentation Every pixel in a picture is given a name during semantic segmentation, allowing for a fine-grained interpretation of the picture's contents. Applications like autonomous navigation, image processing for medical purposes, and visual interpretation can all benefit from this.

3.5 Natural Language Processing

NLP is concerned with the interface between computer and human language. It includes producing human language in a way that computers can grasp, interpret, and analyze [21]. Machine translation, sentiment assessment, text summarization, and automatic speech recognition are all made possible by NLP approaches.

Text Processing Text processing methods involve jobs like tokenization, part-of-speech tagging, named entity identification, and sentiment evaluation. These methods aid in the extraction of useful data from textual material and facilitate additional analysis.

Speech Recognition Speech recognition methods translate spoken words into printed text. To properly recognize and reproduce speech, such techniques make use of methods that include acoustic modeling, linguistic modeling, and sequence-to-sequence modeling.

Sentiment Analysis Sentiment analysis attempts to discern the sentiment or emotion represented in a piece of text. It is employed in analyzing opinions, social media sentiment evaluation, and consumer feedback. Lexical-based methods, ML, and DL models are methods of sentiment analysis.

By studying the principles of AI, such as ML, DL, CV, and NLP, we may dig deeper into the potential uses of AI in autonomous vehicles and investigate how these advancements enable enhanced functions and abilities in the automotive sector.

4 Security Aspects of Hardware Design for Smart Vehicles

Designing intelligent vehicles with AI systems involves a thorough examination of many different aspects [25]. These factors include connection, security and privacy concerns, physical specifications, and software design. These design elements are carefully investigated in this section (Fig. 3).

4.1 Hardware Requirements

Smart vehicles must utilize a variety of physical components to allow capabilities for AI and conduct complicated operations. Among the necessary hardware specifications are the following:

Sensors and Perception Systems: Numerous sensors are included in intelligent vehicles, including cameras, radar, ultrasonic, and LiDAR (light detection and ranging) sensors. These sensors record information about the immediate environment of the vehicle, including the availability of objects, the state of the

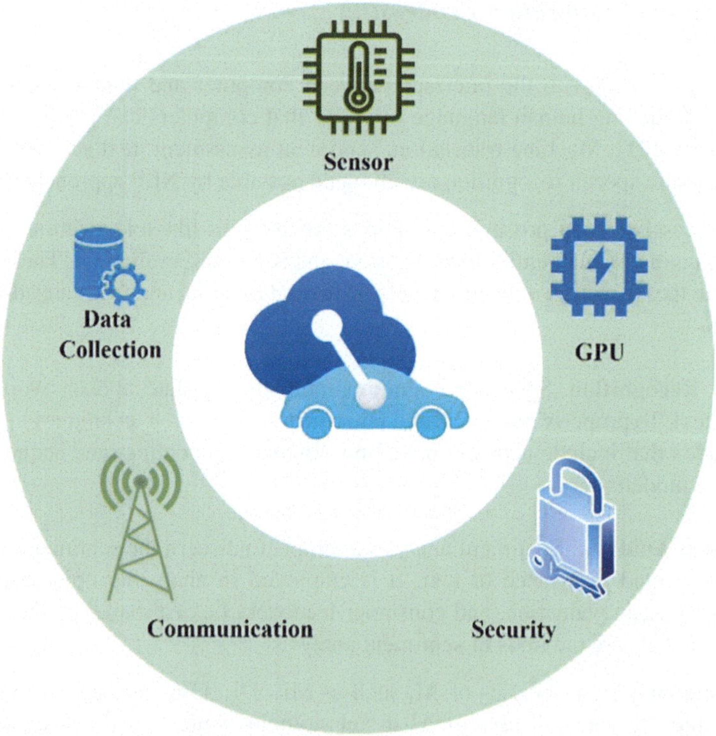

Fig. 3 Hardware design considerations for smart vehicles

roadway, and the movement of traffic. For correct perception and comprehension of the surroundings, high-quality and precise sensors are essential [26].

Processing Units and Architectures: To address the real-time computing demands of AI systems, robust computational units are required. AI computations are frequently accelerated using graphics processing units (GPUs) and specialized AI accelerators (such as TPUs). To fulfill the requirements of AI workloads in automated vehicles, the design of the computing units, including the number of cores, storage space, and efficacy, should also be optimized [27].

4.2 Software Architecture

The software framework of intelligent vehicles is crucial for allowing AI features [25, 26]. Key variables include the following:

Data Collection and Preprocessing: To gather and preserve information from numerous sensors and sources inside the vehicle, reliable systems are needed. Furthermore, the gathered data are processed, such as by cleaning, scaling, and normalization, and transformed into an arrangement compatible with AI models using data preparation methods.

Model Training and Deployment: The offline modeling and online deployment characteristics are needed for AI technology in automated vehicles. The software design should be enabled by training the AI algorithms on large-scale databases, and this is frequently done in cloud or edge computation settings. The algorithms must be effectively implemented and incorporated into the vehicle's onboard technologies after training.

Real-Time Decision-Making: Intelligent vehicles must make real-time observations using information gathered from sensors and AI algorithms. The software design ought to make it easier to integrate decision-making models that can rapidly and accurately evaluate incoming information, analyze the context, and make the right conclusions.

4.3 Connectivity and Communication

Intelligent vehicles depend on a variety of communication tools that allow communication and knowledge sharing across the vehicles, infrastructure, and external frameworks [28]. The main variables include the following:

Vehicle-to-Vehicle (V2V) Communication: Vehicles may communicate with one another via V2V, which allows cooperative actions including platooning, congestion optimization, and prevention of accidents. For smooth V2V conversations, the design should provide dependable and low-latency characteristic protocols for communication.

Vehicle-to-Infrastructure (V2I) Communication: V2I communication entails the transmission of data among vehicles and infrastructure around them, such as traffic lights, roadway markings, and roadway surveillance devices. It makes it possible for cutting-edge features like smart routing, roadway signal prioritization, and real-time updates on traffic conditions. The architecture ought to oblige the incorporation of V2I messaging conventions and protocols.

Vehicle-to-Everything (V2X) Communication: V2V, V2I, and communications with other road users, including bicycles, walkers, and bicyclists, are all included in V2X communication. It makes it possible to drive more safely, with awareness of the situation, and with other people. V2X communication methods should be supported in the development, promising consistent and safe communication routes.

4.4 Safety and Security Considerations

Designing an intelligent vehicle requires careful consideration of security and protection [29, 30]. The following factors should be taken into account while developing AI-powered frameworks:

Functional Safety: To maintain the dependability and integrity of AI systems, intelligent vehicles must follow safety norms and guidelines. To reduce risks and stop collisions, functional security precautions including fail-safe systems, redundant systems, and sturdy error handling should be used.

Cybersecurity: Intelligent automobiles are vulnerable to risks associated with cybersecurity as they grow increasingly interconnected. For automated vehicles to be protected from hostile assaults and to guarantee the privacy of sensitive data, robust safety precautions are required. These measures include encrypted communication protocols, authentication processes, data encryption, and intrusion detection devices.

Engineers may construct intelligent automobiles that efficiently use AI systems by carefully addressing hardware standards, software design, connectivity, and secrecy and privacy concerns. These elements must be considered to create a smooth and safe driving experience, enable enhanced features, and ensure the safety of travelers and other passengers on the road.

Hardware Security The security of the hardware components is crucial for the secure and reliable functioning of intelligent vehicles. Hardware security procedures are intended to safeguard the automobile's hardware parts against illegal access, manipulation, and malicious assaults [31–33]. The different aspects of hardware security in smart vehicles are shown in Fig. 4.

Key factors of hardware security are as follows:

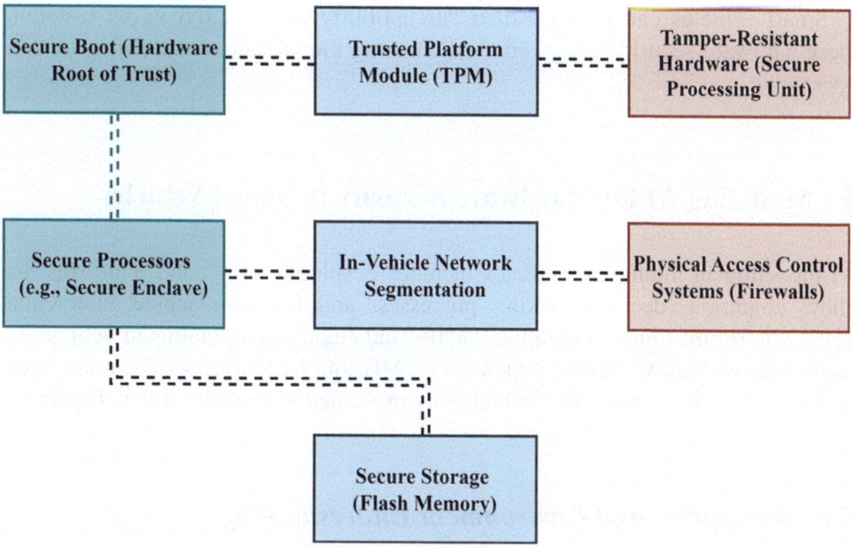

Fig. 4 Components of hardware security in smart vehicles

Secure Boot Mechanism: Putting in place a secure boot procedure guarantees that, when a vehicle starts up, only trustworthy software is loaded and run. This prevents unapproved software from affecting the vehicle's functioning.

Hardware Authentication: By utilizing methods like cryptographic keys kept in encrypted hardware modules, hardware authentication assures that only authorized hardware parts can speak to one another. This stops fraudulent activity and unauthorized access.

Secure Communication Channels: Data sent among hardware elements, such as controllers, gauges, sensors, and actuators, are protected by encryption and shielded from manipulation and espionage when secure communication routes are established.

Equipment Tamper Resistance: Developing tamper-resistant hardware components assists in preventing physical assaults including attempts to access sensitive information or change crucial components. This can be accomplished by utilizing protective enclosures, tamper-evident seals, and dedicated security chips.

Safe Data Storage: Protecting critical information from unwanted access and manipulation is achieved by storing it in hardware-encrypted storage media. This includes user data and encrypted keys.

Hardware Integrity Monitoring: By incorporating hardware integrity monitoring capabilities, the vehicle may identify and react to unlawful changes to its physical elements. This may include looking for alterations to hardware setup or functionality that could signal a safety compromise.

Smart vehicles can improve their susceptibility to cyber threats by installing these hardware security safeguards and ensuring the protection and safety of their operations.

5 Modeling AI for Hardware Security in Smart Vehicles

AI modeling in intelligent vehicles includes developing tools and platforms that allow cognition, decision-making processes, and human–machine interaction. These algorithms improve the functionality and cognition of intelligent vehicles by employing various AI methods, such as CV, ML, and NLP [33, 34]. The main facets of developing AI in automated vehicles are examined in this part of the chapter.

5.1 Perception and Environment Understanding

Perception is essential in intelligent vehicles because it allows them to comprehend and comprehend their surroundings [35]. Prominent modeling strategies for perception involve the following (Fig. 5):

Object Detection and Tracking: Algorithms for object detection and tracking are created to find and recognize objects in the area surrounding the vehicle. These frameworks include object identification methods like the YOLO and SSD as well as CV methods like CNNs that can recognize and monitor people, cars, traffic signs, and numerous other important items.

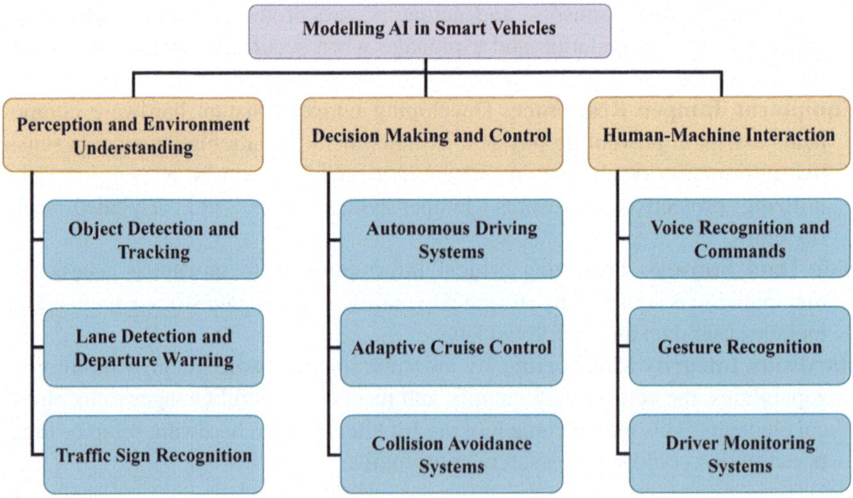

Fig. 5 Modeling AI in smart vehicles

Lane Detection and Departure Warning: Lane markers on the road are identified using lane identification models by analyzing the acquired images or sensor information. These algorithms support the vehicle's ability to maintain position in its allotted lane and offer alerts or mitigation measures if it veers off course.

Traffic Sign Recognition: Algorithms for recognizing traffic signals analyze roadway signs and send information to the system of the vehicle. These frameworks can recognize and comprehend various roadway markings, such as restricting speeds, halt signs, and roadway signals, using CV methods like image categorization or object identification.

5.2 Decision-Making and Control

Smart vehicles may make judgments in real time based on their perception of their environment thanks to AI algorithms for decision-making and control [36]. Key modeling strategies include the following:

Autonomous Driving Systems: AI algorithms are used by self-driving vehicles to interpret sensory information, create itineraries, and manage the movement of vehicles. To steer the car securely and efficiently, these mechanisms use methods including integrating sensors, route planning, and control systems.

Adaptive Cruise Control: Adaptive cruise control (ACC) systems automatically modify the speed of vehicles to keep a secure distance from the vehicle ahead. AI systems for ACC employ sensor information and ML methods to anticipate the actions of other vehicles and alter the speed and acceleration of vehicles automatically.

Collision Avoidance Systems: AI algorithms are used by collision prevention devices to identify and anticipate probable accidents and take appropriate precautions. To assess the likelihood of an accident and launch the proper responses, such as alert notifications or automated braking, these algorithms analyze the surroundings, including the locations and speeds of nearby objects.

5.3 Human–Machine Interaction

The smooth exchange of information among the vehicle and its occupants is made possible by AI algorithms for human–machine communication [33, 35]. Prominent modeling methods involve the following:

Voice Recognition and Commands: Vehicle passengers can perform several tasks without using their hands by voicing commands, which voice recognition models can translate. To correctly comprehend and carry out voice instructions, these AI systems employ methods like automated speech recognition (ASR) and NLP.

Gesture Recognition: With the use of gesture recognition frameworks, a car can comprehend a passenger's hand or body movements. These systems recognize and comprehend certain motions using CV techniques like DL and gesture recognition techniques, enabling passengers to communicate with the vehicle straightforwardly.

Driver Monitoring Systems: Driver behavior, such as movement of the eyes, head posture, and attention, is analyzed by driver surveillance systems using AI methods. These systems can identify symptoms of fatigue, diversion, or impairment and may send warnings or take other steps to protect the safety of the motorist.

Intelligent vehicles may improve their grasp of the surroundings, make cognitive judgments, and offer a smooth and engaging experience for the occupants by utilizing AI systems for cognition, decision-making processes, and human–machine interaction. These models use sophisticated features and help convert cars into intelligent, autonomous, and networked systems by using CV, ML, and NLP methods.

6 Applications of AI in Hardware Security of Smart Vehicles

Automated vehicles utilize AI in several ways, altering the automotive sector and reshaping transportation. Intelligent automobiles can improve security, reliability, and consumer satisfaction by utilizing AI tools including ML, CV, and NLP [37]. This section examines some of the most significant uses of AI in automated automobiles.

6.1 Autonomous Driving

One of the most well-known uses of AI in intelligent vehicles is autonomous driving. Vehicles can now perceive their environment, decipher sensor information, make instantaneous decisions, and maneuver on their thanks to AI systems. Automation levels vary from driver-aid technologies like ACC and lane-keeping assistance to fully autonomous automobiles that can operate without any human intervention [18]. To ensure safety and efficacy in self-driving vehicles, AI-based solutions are used, such as the fusion of sensors, route planning, and control systems (Fig. 6).

Level 1 to Level 5 Automation Self-driving vehicles are divided into five tiers by the Society of Automotive Engineers (SAE). Level 1 consists of driving assistance features including ACC. Partial automation is a feature of level 2; under certain circumstances, the vehicle may concurrently regulate steering, braking, and accelerating. Conditional automation is used at Level 3, where the vehicular system is capable of performing the majority of driving activities but occasionally needs assistance from a person. Level 4 denotes high automation, where the car can drive

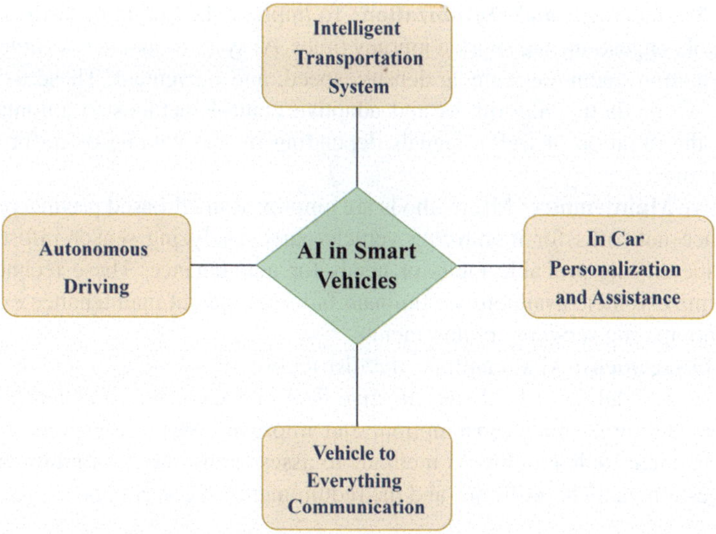

Fig. 6 Applications of AI in hardware security of smart vehicles

itself in particular situations or circumstances. Level 5 is the highest level of automation, where the vehicle is completely driverless and can drive itself in any situation.

Mapping and Localization High-definition maps are produced by AI and used by self-driving vehicles for accurate positioning and navigation. To create comprehensive maps that include roadway limits, road markings, traffic signs, and landmarks, mapping algorithms process sensor data. Self-driving vehicles use such maps to localize themselves precisely and plot the best pathways.

Path Planning and Trajectory Prediction AI-driven routing methods calculate the best route for a vehicle to take to get to its location while taking into factors like traffic norms, congestion, and barriers. To forecast the motion of other vehicles, pedestrians, and objects, trajectory prediction systems employ AI approaches. This enables self-driving vehicles to make wise decisions and plan their paths appropriately.

6.2 Intelligent Transportation Systems

Intelligent vehicular transportation systems (ITS), which optimize congestion management, boost efficacy, and improve the entire travel experience, are developed in large part thanks to AI [36]. A few significant uses are as follows:

Traffic Management and Optimization: To improve the timing of traffic signals, control congestion, and shorten journey times, AI systems assess real-time traffic information, including vehicle density, speed, and movement. These technologies use predictive algorithms and adaptive control methods to automatically alter the duration of traffic signals depending on the volume of traffic at any given time.

Predictive Maintenance: ML methods are employed in AI-based predictive maintenance networks for monitoring vehicle parts, analyzing sensor information, and identifying probable faults or needs for maintenance. These technologies maximize vehicle availability, eliminate failures, and cut maintenance expenses by anticipating servicing requirements.

Fleet Management: AI streamlines the distribution of resources, route design, as well as scheduling to facilitate effective fleet management. To enhance performance, minimize fuel consumption, and improve consumer experience, fleet management tools employ AI methods to assess real-time information such as congestion, vehicle positions, and the requirements of consumers.

6.3 In-car Personalization and Assistance

Intelligent assistance for drivers and customized in-car experiences are made possible by AI [25]. Several uses are as follows:

Infotainment Systems: Vehicle occupants receive individualized communications, entertainment, and navigational features through infotainment systems that are powered by AI. These platforms combine ML and NLP to recognize consumer preferences, suggest content, and support voice-activated engagements.

Voice-Controlled Assistants: Intelligent virtual assistants powered by AI allow users to communicate with the vehicle via natural language requests. Automatic speech recognition (ASR) and natural language understanding (NLU) methods are employed by virtual assistants to analyze verbal instructions, respond to inquiries, give directions, and manage numerous vehicle operations.

Driver Behavior Analysis: To improve safety and offer individualized feedback, AI systems assess driver actions, including driving patterns, alertness, and degree of weariness. Driver behavior analysis tools can identify risky driving patterns, send out alarms, and help drivers in developing safer driving practices.

6.4 Vehicle-to-Everything (V2X) Communication

Vehicles may interact with physical facilities, external systems, as well as other vehicles using V2X communication [25, 36]. AI is used to interpret and evaluate V2X information, enabling programs as follows:

Cooperative Collision Avoidance: Collaborative collision prevention is made possible through V2X communication, which allows vehicles to exchange real-time data regarding their positions, velocity, and intents. These data are processed by AI systems to identify probable accidents and aid in taking precautions.

Traffic Signal Prioritization: Vehicles can get priority access to roadway signals by communicating with traffic control systems, which eases congestion and improves the flow of traffic. Using real-time traffic information analysis and considerations for vehicle usage and traffic circumstances, AI systems optimize signal prioritizing.

Infrastructure-to-Vehicle (I2V) Safety Warnings: Infrastructure networks can provide security notifications to moving vehicles regarding threats on the road, work zones, or emergency circumstances thanks to V2X connectivity. These signals are processed by AI systems, which help in providing timely notifications and direction to the occupants of moving vehicles.

The automotive industry is being transformed, and transportation itself is being redefined, by the numerous and ever-evolving applications of AI in automated vehicles. Advanced automobiles improve safety, improve congestion control, offer tailored experiences, and pave the path for the development of automated and networked transportation thanks to artificial intelligence (AI) innovations.

7 Challenges and Future Directions

The incorporation of AI in the hardware security of intelligent vehicles involves many hurdles as well as intriguing future modes of transportation opportunities. To the fullest extent possible, AI in smart vehicles must be able to overcome these obstacles and navigate the way forward [38–40]. This section examines a few of the main issues and potential paths for AI in connected and autonomous automobiles.

7.1 Ethical Considerations

Ethical issues must be addressed as AI technology becomes more common in automated vehicles. Concerns about legitimacy and accountability for AI-driven judgments are raised, particularly in dire circumstances. To foster trust and address societal issues, it is crucial that AI models ensure openness, justice, and ethical decision-making.

7.2 Regulatory and Legal Frameworks

The creation of thorough regulatory and legislative structures is necessary for the implementation of AI in autonomous vehicles. These frameworks should cover topics including accountability, secrecy, cybersecurity, and preventive regulations. The appropriate research, implementation, and use of AI methods for autonomous vehicles are going to be ensured by explicit laws and policies.

7.3 Data Privacy and Security

Smart vehicles generate and gather enormous volumes of information, including critical sensor information as well as individual information. It is essential to guarantee the confidentiality and safety of sensitive information. To protect against cyber risks and protect the confidentiality of motorist occupants, restricted access controls, encrypted communication protocols, and strong safety precautions for information need to be put in place.

7.4 Scalability and Integration

It will become more difficult as AI technology develops to guarantee the scalability and smooth execution of AI systems in automated vehicles. Real-time processing and the computational demands of AI algorithms require robust hardware and well-designed software systems. To ensure the cost-effective implementation of AI in intelligent vehicles, efficient integration of AI technologies with already existing vehicle systems and parts is going to be essential.

7.5 Human Trust and Acceptance

Successful deployment of AI-driven smart vehicles depends on establishing consumer acceptance and confidence. Gaining trust among consumers and promoting acceptance of AI technology in autonomous vehicles will depend on providing explicit justifications for AI-driven choices, guaranteeing transparency in the process of making choices, and solving user concerns about security as well as management.

7.6 *Future Prospects of AI in Secure Hardware of Smart Vehicles*

Moving ahead, several interesting developments for AI in smart vehicles might be projected:

Advanced Autonomous Driving: Hardware security in advanced autonomous vehicles will keep evolving as a result of improvements in AI systems. Further studies should focus on enhancing sensing abilities, enhancing decision-making computations, and dealing with difficult situations like construction sites and severe weather. Complete autonomy and other greater degrees of automation will transform transportation and improve its effectiveness and security.

AI-Enabled Cooperative Systems: There is a lot of promise for AI-driven collaborative systems that allow for seamless interaction and collaboration between infrastructure, automobiles, and pedestrians. Through coordinated collision prevention, such technologies can improve traffic flow, improve security, and provide efficient power control in public transit systems.

Context-Aware Personalization: Based on individual preferences, demands, and circumstances, AI systems will increasingly tailor the in-vehicle experience. A personalized and interesting environment will be created for passengers in intelligent vehicles by offering customized suggestions for amusement, navigation, and assistance.

Intelligent Mobility Services: Smart transport systems will eventually be developed thanks to the incorporation of AI in the hardware security of automated vehicles. On-demand sharing rides, flexible routing, improved fleet management, and unified multifaceted transportation solutions are a few examples of these offerings. Mobility solutions powered by AI can transform urban transportation while reducing traffic and its ecological effects.

Despite the great possibilities that AI offers for smart automobiles, issues with ethics, legislation, confidentiality, flexibility, approval, and confidence must be resolved. By addressing these barriers, we can create an ecosystem where AI-driven autonomous vehicles improve consumer experience, security, and productivity while redefining the way people travel today and in the near future.

8 Conclusion

The modern age of transportation has begun with the emergence of AI methods in intelligent vehicles, which has transformed the automotive sector and the way we move. Intelligent decision-making, improved hardware security, productivity, and consumer satisfaction are all made possible by AI in autonomous vehicles. Through this study, we have looked at the foundations, design factors, modeling methods, and potential uses of AI in the hardware security of automated vehicles. ML, DL,

CV, and NLP are the key elements of AI that serve as a foundation of automated systems to build security aspects in smart automobiles. Vehicles can acquire knowledge from information, comprehend imagery data, incorporate natural language instructions, and come up with sensible choices considering uncertain circumstances in security aspects thanks to these innovations.

Discovering the possibilities of AI in the hardware security of automated vehicles depends critically on design factors. Real-time AI computations need an intelligent selection of hardware specifications, comprising sensors and computational units. The procedures of gathering information, preparing this, training models, and executing decisions must all be made easier by software design. Collaborative behaviors are made possible by communication and connectivity innovations, such as V2V and V2I interactions, which improve traffic control and security.

Intelligent vehicle modeling concerns various security aspects of vehicle hardware for perception, comprehension of the surroundings, management, and human–machine interaction. To identify and monitor objects, estimate the best routes, and enable autonomous driving, AI models examine data collected by sensors. Human–machine interactions identify movements and comprehend spoken instructions, boosting the in-vehicle experience. Automated driving, intelligent public transit, in-car personalization, and V2X connectivity are just a few of the numerous uses of AI in the hardware security of intelligent vehicles. These programs alter how we communicate with vehicles, improve congestion control, tailor our experiences, and increase security.

Nevertheless, incorporating hardware security utilizing AI into automated vehicles also presents issues that must be resolved. To enable the appropriate and secure use of AI technology, adequate attention must be given to moral concerns, legislation, data security and privacy, flexibility, human confidence, and acceptability. The use of AI in autonomous vehicles offers enormous potential in the future. The next-generation transport system is going to be shaped by developments of secure hardware in self-driving vehicles, collaborative systems, context-aware personalization, and smart transport services, all of which will increase security, effectiveness, and sustainability.

To summarize, AI systems in automated vehicles are transforming the automotive sector by allowing vehicles to become smart, autonomous, and networked. The future of transportation can be made more secure, efficient, and completely connected to our lives by utilizing AI tools. We shall enter a transformational era of transportation if we solve the challenges and accept the potential of AI in hardware security aspects of automated vehicles.

References

1. Bendiab, G., Hameurlaine, A., Germanos, G., Kolokotronis, N., & Shiaeles, S. (2023). Autonomous vehicles security: Challenges and solutions using blockchain and artificial intelligence. *IEEE Transactions on Intelligent Transportation Systems, 24*(4), 3614–3637. https://doi.org/10.1109/TITS.2023.3236274
2. Padmanaban, S., Samavat, T., Nasab, M. A., Nasab, M. A., Zand, M., & Nikokar, F. (2023). Electric vehicles and IoT in smart cities. In *Artificial intelligence-based smart power systems* (pp. 273–290). Wiley. https://doi.org/10.1002/9781119893998.ch14
3. Cao, D., Wang, X., Li, L., Lv, C., Na, X., Xing, Y., Li, X., Li, Y., Chen, Y., & Wang, F. Y. (2022). Future directions of intelligent vehicles: Potentials, possibilities, and perspectives. *IEEE Transactions on Intelligent Vehicles, 7*(1), 7–10. https://doi.org/10.1109/TIV.2022.3157049
4. Bathla, G., Bhadane, K., Singh, R. K., Kumar, R., Aluvalu, R., Krishnamurthi, R., Kumar, A., Thakur, R. N., & Basheer, S. (2022). Autonomous vehicles and intelligent automation: Applications, challenges, and opportunities. *Mobile Information Systems*. https://doi.org/10.1155/2022/7632892
5. Whig, P., Velu, A., Nadikattu, R. R., & Alkali, Y. J. (2024). Role of AI and IoT in intelligent transportation. In *artificial intelligence for future intelligent transportation* (pp. 199–220). Apple Academic Press. https://doi.org/10.1201/9781003408468
6. Alshdadi, A. A. (2021). Cyber-physical system with IoT-based smart vehicles. *Soft Computing, 25*(18), 12261–12273. https://doi.org/10.1007/s00500-021-05908-w
7. Pali, I., Amin, R., & Abdussami, M. (2024). Autonomous vehicle security: Current survey and future research challenges. *Security and Privacy, 7*(3), e367. https://doi.org/10.1002/spy2.367
8. Alalwany, E., & Mahgoub, I. (2024). Security and trust management in the Internet of Vehicles (IoV): Challenges and machine learning solutions. *Sensors, 24*(2), 368. https://doi.org/10.3390/s24020368
9. Muslam, M. M. A. (2024). Enhancing security in vehicle-to-vehicle communication: A comprehensive review of protocols and techniques. *Vehicles, 6*(1), 450–467. https://doi.org/10.3390/vehicles6010020
10. Naeem, A. B., Soomro, A. M., Saim, H. M., & Malik, H. (2024). Smart road management system for prioritized autonomous vehicles under vehicle-to-everything (V2X) communication. *Multimedia Tools and Applications, 83*(14), 41637–41654. https://doi.org/10.1007/s11042-023-16950-1
11. Alqahtani, H., & Kumar, G. (2024). Machine learning for enhancing transportation security: A comprehensive analysis of electric and flying vehicle systems. *Engineering Applications of Artificial Intelligence, 129*, 107667. https://doi.org/10.1016/j.engappai.2023.107667
12. Ramalakshmi, K., Ganesh, S., & Krishna Kumari, L. (2024). Security and privacy issues of AI in autonomous vehicles. In *Artificial intelligence for autonomous vehicles* (pp. 229–246). https://doi.org/10.1002/9781119847656.ch11
13. Banafshehvaragh, S. T., & Rahmani, A. M. (2023). Intrusion, anomaly, and attack detection in smart vehicles. *Microprocessors and Microsystems, 96*, 104726. https://doi.org/10.1016/j.micpro.2022.104726
14. Alfrink, K., Keller, I., Doorn, N., & Kortuem, G. (2023). Tensions in transparent urban AI: designing a smart electric vehicle charge point. *AI & Society, 38*(3), 1049–1065. https://doi.org/10.1007/s00146-022-01436-9
15. Karnati, A., & Mehta, D. (2022). Artificial Intelligence in Self Driving Cars: Applications, Implications and Challenges. *Ushus Journal of Business Management, 21*(4), 1–28. https://doi.org/10.12725/ujbm/61.0
16. Kaur, P., Kumar, Y., & Gupta, S. (2022). Artificial intelligence techniques for the recognition of multi-plate multi-vehicle tracking systems: A systematic review. *Archives of Computational Methods in Engineering, 29*(7), 4897–4914. https://doi.org/10.1007/s11831-022-09753-4

17. Di, X., & Shi, R. (2021). A survey on autonomous vehicle control in the era of mixed-autonomy: From physics-based to AI-guided driving policy learning. *Transportation Research Part C: Emerging Technologies, 125*, 103008. https://doi.org/10.1016/j.trc.2021.103008
18. Biggi, G., & Stilgoe, J. (2021). Artificial intelligence in self-driving cars research and innovation: A scientometric and bibliometric analysis. *SSRN, 3829897*. https://doi.org/10.2139/ssrn.3829897
19. Ma, Y., Wang, Z., Yang, H., & Yang, L. (2020). Artificial intelligence applications in the development of autonomous vehicles: A survey. *IEEE/CAA Journal of Automatica Sinica, 7*(2), 315–329. https://doi.org/10.1109/JAS.2020.1003021
20. Thadeshwar, H., Shah, V., Jain, M., Chaudhari, R., & Badgujar, V. (2020). Artificial intelligence based self-driving car. In *2020 4th International Conference on Computer, Communication and Signal Processing (ICCCSP)* (pp. 1–5). IEEE. https://doi.org/10.1109/ICCCSP49186.2020.9315223
21. Rodriguez-Fernandez, V., & Camacho, D. (2024). Recent trends and advances in machine learning challenges and applications for industry 4.0. *Expert Systems, 41*(2), e13506. https://doi.org/10.1111/exsy.13506
22. Chauhan, N. K., & Singh, K. (2018). A review on conventional machine learning vs deep learning. In *2018 International conference on computing, power and communication technologies (GUCON)* (pp. 347–352). IEEE. https://doi.org/10.1109/GUCON.2018.8675097
23. Chauhan, N. K., & Singh, K. (2021). Impact of variation in number of channels in CNN classification model for cervical cancer detection. In *2021 9th international conference on reliability, infocom technologies and optimization (Trends and future directions) (ICRITO)* (pp. 1–6). IEEE. https://doi.org/10.1109/ICRITO51393.2021.9596366
24. Mishra, A., & Kim, S. (2023). Irregular situations in real-world intelligent systems. *Advanced Computing, 134*, 1–31. https://doi.org/10.1016/bs.adcom.2023.04.006
25. Abbasi, S., & Rahmani, A. M. (2023). Artificial intelligence and software modeling approaches in autonomous vehicles for safety management: A systematic review. *Information, 14*(10), 555. https://doi.org/10.3390/info14100555
26. Kiran, B. R., Sobh, I., Talpaert, V., Mannion, P., Al Sallab, A. A., Yogamani, S., & Pérez, P. (2021). Deep reinforcement learning for autonomous driving: A survey. *IEEE Transactions on Intelligent Transportation Systems, 23*(6), 4909–4926. https://doi.org/10.1109/TITS.2021.3054625
27. Gao, X., & Bian, X. (2021). Autonomous driving of vehicles based on artificial intelligence. *Journal of Intelligent Fuzzy Systems, 41*(4), 4955–4964. https://doi.org/10.3233/JIFS-189982
28. Khayyam, H., Javadi, B., Jalili, M., & Jazar, R. N. (2020). Artificial intelligence and internet of things for autonomous vehicles. In *Nonlinear approaches in engineering applications: Automotive applications of engineering problems* (pp. 39–68). Springer. https://doi.org/10.1007/978-3-030-18963-1_2
29. Mishra, A., Kim, J., Cha, J., Kim, D., & Kim, S. (2021). Authorized traffic controller hand gesture recognition for situation-aware autonomous driving. *Sensors, 21*(23), 7914. https://doi.org/10.3390/s21237914
30. Rimal, B. P., Kong, C., Poudel, B., Wang, Y., & Shahi, P. (2022). Smart electric vehicle charging in the era of internet of vehicles, emerging trends, and open issues. *Energies, 15*(5), 1908. https://doi.org/10.3390/en15051908
31. Yadav, P., Mishra, A., & Kim, S. (2023). A comprehensive survey on multi-agent reinforcement learning for connected and automated vehicles. *Sensors, 23*, 4710. https://doi.org/10.3390/s23104710
32. Mishra, A., Lee, S., Kim, D., & Kim, S. (2022). In-cabin monitoring system for autonomous vehicles. *Sensors, 22*(12), 4360. https://doi.org/10.3390/s22124360
33. Wang, Z., Zhan, J., Duan, C., Guan, X., Lu, P., & Yang, K. (2022). A review of vehicle detection techniques for intelligent vehicles. *IEEE Transactions on Neural Networks and Learning Systems, 34*(8), 3811–3831. https://doi.org/10.1109/TNNLS.2021.3128968

34. Samanta, D. P. S., Patnaik, B. C. M., Satpathy, I., & Akbar, J. (2023). Analyzing bibliometrics of AI and smart vehicles. In *Handbook of research on AI-based technologies and applications in the Era of the Metaverse* (pp. 359–380). IGI Global. https://doi.org/10.4018/978-1-6684-8851-5.ch019
35. Partovi, Z., Zarei, M., & Rahmani, A. M. (2023). Data-centric approaches in the internet of vehicles: A systematic review on techniques, open issues, and future directions. *International Journal of Communication Systems, 36*(3), e5383. https://doi.org/10.1002/dac.5383
36. Mishra, A., Cha, J., & Kim, S. (2022). Privacy-preserved in-cabin monitoring system for autonomous vehicles. *Computational Intelligence and Neuroscience*. https://doi.org/10.1155/2022/5389359
37. Parekh, D., Poddar, N., Rajpurkar, A., Chahal, M., Kumar, N., Joshi, G. P., & Cho, W. (2022). A review on autonomous vehicles: Progress, methods and challenges. *Electronics, 11*, 2162. https://doi.org/10.3390/electronics11142162
38. Xiong, R., Kim, J., Shen, W., Lv, C., Li, H., Zhu, X., Zhao, W., Gao, B., Guo, H., Zhang, C., & Sun, F. (2022). Key technologies for electric vehicles. *Green Energy and Intelligent Transportation, 1*(2), 100041. https://doi.org/10.1016/j.geits.2022.100041
39. Al-Hanahi, B., Ahmad, I., Habibi, D., & Masoum, M. A. (2021). Charging infrastructure for commercial electric vehicles: Challenges and future works. *IEEE Access, 9*, 121476–121492. https://doi.org/10.1109/ACCESS.2021.3108817
40. Onat, N. C., & Kucukvar, M. (2022). A systematic review on sustainability assessment of electric vehicles: Knowledge gaps and future perspectives. *Environmental Impact Assessment Review, 97*, 106867. https://doi.org/10.1016/j.eiar.2022.106867

Blockchain for Hardware Security and Trust

Pawan Whig, Iti Batra, Nikhitha Yathiraju, and Seema Nath Jain

1 Introduction

The digital landscape of the twenty-first century is marked by an explosive proliferation of embedded systems, powering a diverse array of devices, from our smartphones and home appliances to industrial machinery and critical infrastructure components. These embedded systems represent the backbone of modern life, facilitating seamless communication, automation, and control. However, this ubiquity also makes them prime targets for cyberattacks, emphasizing the pressing need for robust hardware security measures. In this pivotal chapter, we embark on a captivating journey into the intricate intersection of blockchain technology and hardware security. As we delve into this dynamic realm, we unveil a symbiotic relationship between two cutting-edge domains, poised to revolutionize the way we safeguard and trust our digital infrastructure.

The significance of security in embedded systems cannot be overstated. These systems are integral to the functioning of our interconnected world, influencing not only the convenience of our daily lives but also the stability of critical services and infrastructures, including health care, energy, transportation, and finance. Consequently, their vulnerability to malicious attacks poses profound threats to society, both in terms of economic impact and public safety. Traditionally, hardware security measures have focused on safeguarding these embedded systems through methods such as secure boot processes, hardware root of trust, and tamper-resistant

P. Whig (✉) · I. Batra
Vivekananda Institute of Professional Studies – Technical Campus, New Delhi, India

N. Yathiraju
University of The Cumberlands, Williamsburg, KY, USA

S. N. Jain
Vivekananda College, Delhi University, Delhi, India
e-mail: seemanath@vivekanand.du.ac.in

© The Author(s), under exclusive license to Springer Nature Switzerland AG 2025
A. Mishra et al. (eds.), *Hardware Security: Challenges and Solutions*,
https://doi.org/10.1007/978-3-031-81213-2_2

modules. While effective, these measures often operate within isolated silos, and the complexity of modern hardware ecosystems demands a more holistic approach to security and trust. Blockchain has evolved into a multifaceted tool that transcends financial applications, offering unparalleled security features, decentralization, immutability, and consensus mechanisms that can be harnessed to fortify hardware systems. Its potential to create transparent, tamper-resistant, and auditable ledgers aligns seamlessly with the objectives of hardware security. Our exploration commences with a foundational understanding of hardware security principles. We delve into the fundamentals, examining the intricacies of hardware vulnerabilities and threats that challenge the integrity of embedded systems. Secure boot processes and hardware roots of trust are dissected, shedding light on the critical role they play in establishing a trustworthy foundation. As we progress, we uncover how blockchain technology can be integrated into embedded systems to enhance their security and trustworthiness. Real-world use cases exemplify the benefits of this integration, from securing Internet of Things (IoT) devices against evolving threats to bolstering supply chain security through transparent and immutable record-keeping.

The chapter also navigates the terrain of privacy-preserving protocols and cryptographic innovations, essential components in the arsenal against digital threats. We explore cutting-edge techniques that enable hardware systems to maintain data confidentiality while interacting with a decentralized blockchain network. Vulnerability mitigation takes center stage as we discuss strategies to identify, analyze, and address hardware vulnerabilities. Ensuring the trustworthiness of hardware systems is a cornerstone of this discussion, emphasizing the importance of comprehensive security measures. Looking ahead, we contemplate the future directions and emerging trends at the confluence of blockchain and hardware security. Ethical considerations and regulatory implications are brought to the forefront, guiding us in addressing the complex ethical dilemmas and compliance challenges that this fusion of technologies presents. In essence, this chapter is our gateway to a realm where blockchain and hardware security converge, reshaping the landscape of digital trust. It is a testament to the inexhaustible capacity of human innovation in the face of evolving threats, offering a roadmap for engineers, researchers, and cybersecurity professionals to navigate this transformative landscape with confidence and resilience. Together, we embark on a journey to secure the foundation of our digital world, one embedded system at a time.

1.1 Background and Motivation

The integration of blockchain technology into hardware security represents a pivotal moment in the evolution of cybersecurity. To appreciate the significance of this convergence, it is essential to delve into the background that has led us to this point. In recent years, the proliferation of embedded systems across various industries and the rapid expansion of the Internet of Things (IoT) have exposed a glaring vulnerability in our digital infrastructure. These embedded systems, ranging from medical devices and autonomous vehicles to smart home appliances, are interconnected, often remotely managed, and continuously collect and transmit data. While this connectivity offers unprecedented convenience and efficiency, it also opens a Pandora's box of security challenges. Cyberattacks, ranging from data breaches and device tampering to supply chain infiltrations, have become all too common, threatening the privacy and integrity of both personal and critical data. Traditional hardware security measures have struggled to keep pace with the evolving threat landscape. They often rely on isolated and static security mechanisms that can be exploited or circumvented by determined adversaries. This vulnerability has necessitated a paradigm shift in how we approach hardware security, leading us to explore the innovative integration of blockchain technology.

Blockchain's emergence as a transformative force in the world of digital finance has showcased its remarkable capabilities in ensuring trust, immutability, and decentralized consensus. These very attributes hold immense promise for bolstering hardware security. By harnessing the power of blockchain, we can establish transparent and tamper-resistant ledgers that verify the authenticity and integrity of hardware components and transactions. This fusion of technologies has the potential to mitigate vulnerabilities, enhance supply chain security, and provide a robust defense against emerging cyber threats. Literature review is shown in Table 1.

1.2 Objectives

This chapter is dedicated to elucidating the intricate relationship between blockchain and hardware security and the transformative impact it can have on the digital landscape. Our objectives are threefold:

1. **Understanding the Synergy**: We aim to provide a comprehensive understanding of how blockchain technology and hardware security intersect and complement each other. By exploring the fundamentals of both domains, readers will gain insights into the synergistic potential that can be harnessed to fortify hardware-based systems [9–17].
2. **Practical Applications**: Through real-world use cases and case studies, we will showcase the practical applications of integrating blockchain into hardware security. These examples will highlight the tangible benefits of this amalgamation, from securing IoT devices to enhancing supply chain transparency [18, 19].
3. **Navigating Challenges and Opportunities**: It is crucial to address the challenges and considerations that arise when implementing blockchain in hardware security. We will delve into the potential obstacles, regulatory considerations, and ethical dilemmas associated with this transformative fusion, providing readers with a holistic view of the landscape.

As we embark on this exploration, readers will be equipped with the knowledge and insights necessary to navigate the exciting and complex terrain of blockchain for

Table 1 Literature review

Reference	Key findings	Research gap
Kulkarni et al. [1]	Blockchain enhances IC supply chain security and trust.	More exploration needed on the scalability of blockchain in large-scale IC supply chains.
Cabrera-Gutiérrez et al. [2]	Blockchain and HSM integration improves security in industrial IoT networks.	Limited focus on the practical implementation challenges and cost implications.
Singh & Michels [3]	Study on Blockchain as a Service (BaaS) provider and trust issues.	Lack of research on BaaS impact in various industries.
Alrubei et al. [4]	Addition of hardware security into IoT blockchain platforms.	Limited discussion on the trade-offs between hardware security and scalability.
Xu et al. [5]	Trustless architecture of blockchain in metaverse applications.	Need for further exploration on real-world use cases and scalability.
Saha et al. [6]	Integration of hardware security into a blockchain-based transactive energy platform.	Limited research on the impact of real-time hardware security in energy systems.
Sklavos et al. [7]	Overview of hardware security and trust.	More recent research needed on the integration of blockchain with hardware security.
Jesus [8]	Exploration of blockchain-enhanced roots-of-trust.	Limited focus on practical applications and industry use cases.
Abed et al. [9]	Review on blockchain and IoT integration from energy, security, and hardware perspectives.	Gaps in understanding the practical challenges of implementing blockchain in IoT.
Islam & Kundu [10]	IoT security, privacy, and trust in home-sharing economy via blockchain.	Need for research on the scalability of blockchain in IoT for home-sharing applications.
Enkhtaivan et al. [11]	Fair anonymous auction scheme using trusted hardware and blockchain.	Limited exploration of real-world applications beyond auctions.
Bentov et al. [12]	Real-time cryptocurrency exchange using trusted hardware.	More research needed on security and regulatory implications.
Gountia [13]	Examination of scalability and security issues in state-of-the-art blockchain.	Need for more recent studies addressing scalability challenges.
Cui et al. [14]	Review of blockchain in IoT trends and challenges.	Limited research on practical challenges and real-world use cases.
Boireau [15]	Focus on securing the blockchain against hackers.	Need for exploration of evolving security threats and countermeasures.
Novotny et al. [16]	Study on permissioned blockchain technologies for academic publishing.	Limited exploration of blockchain applications beyond publishing.

hardware security and trust, ensuring the continued advancement of digital technology with resilience and confidence [20–23].

2 Understanding Hardware Security

Hardware security constitutes the foundation upon which the trustworthiness of embedded systems is built. To grasp the significance of blockchain's integration into this realm, it is crucial to first delve into the fundamentals of hardware security. This section will provide an essential overview of hardware security principles, vulnerabilities, and the core strategies employed to protect against threats [24].

2.1 Fundamentals of Hardware Security

At its core, hardware security encompasses a broad spectrum of practices and measures designed to safeguard the integrity, confidentiality, and availability of hardware components and systems [25–28]. It involves both physical and logical aspects of security, addressing threats that can originate from various sources, including malicious actors, environmental conditions, and manufacturing processes. The fundamental aspects of hardware security include

Physical Security This pertains to safeguarding hardware components from physical tampering, theft, or damage. Measures such as secure enclosures, access controls, and tamper-evident mechanisms are often employed to deter and detect physical attacks.

Authentication Ensuring that only authorized users or devices can access and interact with hardware systems is a cornerstone of security. Authentication mechanisms, such as passwords, biometrics, and cryptographic keys, play a pivotal role in this regard.

Authorization Once authenticated, users or devices may have varying levels of access privileges within the hardware system. Authorization controls dictate what actions and resources are accessible to each entity, mitigating potential misuse or unauthorized access.

Data Integrity Maintaining the integrity of data as it flows through hardware components is essential. Cryptographic techniques, such as digital signatures and message authentication codes, are often employed to detect unauthorized alterations to data.

2.2 Hardware Vulnerabilities and Threats

Hardware, while critical to the functionality of digital systems, is not immune to vulnerabilities and threats. Understanding these vulnerabilities is paramount to devising effective security measures. Some common hardware vulnerabilities are as follows:

Design Flaws: Vulnerabilities can originate from flaws in the hardware's design, such as improper input validation, weak encryption, or insecure default configurations [29–33].

Supply Chain Risks: Malicious actors can exploit vulnerabilities at various stages of the hardware supply chain, from manufacturing to distribution. Counterfeit components, hardware implants, and supply chain compromises pose significant threats.

Physical Attacks: Physical attacks encompass a range of threats, from invasive attacks that involve tampering with hardware components to side-channel attacks that exploit unintended leakage of information, such as power consumption or electromagnetic radiation.

Firmware and Software Vulnerabilities: Vulnerabilities in firmware and software running on hardware components can compromise security. This includes vulnerabilities in BIOS/UEFI firmware, device drivers, and operating systems.

2.3 Role of Blockchain in Hardware Security

As we delve further into this chapter, we will explore how blockchain technology can augment and enhance hardware security measures. Blockchain's inherent properties, including decentralization, immutability, and consensus mechanisms, provide new avenues for addressing these vulnerabilities and bolstering the trustworthiness of embedded systems. By integrating blockchain into hardware security, we unlock novel methods for safeguarding against threats and ensuring the integrity of hardware components and transactions [34].

3 Implementing Blockchain in Embedded Systems

The integration of blockchain technology into embedded systems represents a paradigm shift in hardware security. In this section, we delve into the practical aspects of implementing blockchain within these systems, exploring the challenges and solutions associated with this transformative endeavor.

3.1 Integration Challenges and Solutions

Integrating blockchain into embedded systems presents unique challenges, primarily due to the resource-constrained nature of many embedded devices [35–37]. Blockchain's decentralized nature, which involves maintaining a full ledger across a network of nodes, can strain the limited computational power, memory, and storage capacity of these devices. Some key integration challenges are as follows:

Resource Constraints: Embedded systems often have limited computational resources, making it challenging to participate in blockchain networks that demand significant computational power for tasks like mining or transaction validation.

Storage Limitations: The blockchain ledger can grow substantially over time, consuming a valuable storage space on embedded devices. This can be problematic for devices with limited storage capacity.

Energy Efficiency: Many embedded systems are battery-powered and must operate efficiently to maximize battery life. Blockchain operations can be energy-intensive, posing a challenge for energy-efficient operation.

Solutions to these challenges:

Blockchain Pruning: Implementations that allow for pruning or selective storage of blockchain data can reduce storage requirements on embedded devices.

Lightweight Protocols: Using lightweight blockchain protocols tailored for resource-constrained devices can optimize resource usage while still maintaining the security and trust features of blockchain.

Consensus Mechanisms: Employing energy-efficient consensus mechanisms, such as Proof of Stake (PoS) or delegated Proof of Stake (DPoS), can mitigate the energy consumption issue.

3.2 Hardware Root of Trust and Secure Boot

A critical aspect of implementing blockchain in embedded systems is establishing a secure foundation. This includes ensuring the integrity of the hardware components and the software stack running on them [38, 39]. Two fundamental components in this regard are the Hardware Root of Trust (HRoT) and Secure Boot.

Hardware Root of Trust (HRoT) HRoT is a hardware-based component that provides a foundation of trust for the entire system. It is typically a dedicated hardware module that stores cryptographic keys and performs key management functions. HRoT is crucial for protecting sensitive data and verifying the authenticity of the software that runs on the system.

Secure Boot Secure Boot is a process by which the system verifies the integrity and authenticity of the firmware and software components during the boot process. It ensures that only trusted and digitally signed code is executed, mitigating the risk of unauthorized or malicious software running on the device.

Blockchain technology can enhance HRoT and Secure Boot processes by adding an additional layer of trust through transparent and tamper-resistant ledgers. The blockchain ledger can record and verify the cryptographic signatures of firmware and software components, ensuring that only authorized and unaltered code is executed. By integrating blockchain with HRoT and Secure Boot, embedded systems can achieve a higher level of security and trust, making them resilient to a wide range of threats, including supply chain compromises and unauthorized software modifications [40].

4 Case Studies

To illustrate the practical implications and benefits of integrating blockchain technology into embedded systems, let us delve into real-world case studies. These examples showcase the diverse applications and successes of blockchain in fortifying hardware security.

4.1 Securing IoT Devices with Blockchain

Case Study 1: Protecting Smart Home Devices
In an era of connected homes, smart devices are susceptible to cyber threats. This case study explores how a leading smart home technology provider integrated blockchain into its devices to enhance security. By using blockchain for device authentication and communication, they achieved the following:

Enhanced Authentication: Blockchain's decentralized authentication ensured that only authorized users and devices could access and control smart home devices, thwarting unauthorized access attempts.

Immutable Logs: Transaction records, firmware updates, and device interactions were securely logged on the blockchain. This provided an immutable audit trail, valuable for diagnosing issues and ensuring data integrity.

Supply Chain Verification: Blockchain was utilized to trace the provenance of components in the supply chain, reducing the risk of counterfeit or compromised hardware.

How to protect smart home devices, with a focus on using blockchain technology to enhance security? Protecting smart home devices is crucial as they form a significant part of the Internet of Things (IoT) ecosystem and often handle sensitive data and control over our home, as shown in Fig. 1.

1. Authentication and Authorization

Blockchain Integration Blockchain can be employed to establish secure authentication and authorization mechanisms for smart home devices. Each device can have a unique identity stored on the blockchain, and access permissions can be managed through smart contracts.

User-Friendly Interfaces User-friendly interfaces, such as mobile apps, can interact with the blockchain to manage device access. Users can grant or revoke permissions, ensuring that only authorized users can control their devices.

2. Data Encryption

End-to-End Encryption Implement end-to-end encryption for data transmitted between smart devices and control interfaces. Blockchain can store and manage encryption keys securely, ensuring that only authorized parties can decrypt and access the data.

3. Immutable Device Logs

Blockchain Ledger Store device interaction logs and access history on the blockchain. This provides an immutable and tamper-resistant record of all device activities. Any unauthorized access attempts or tampering will be immediately detectable.

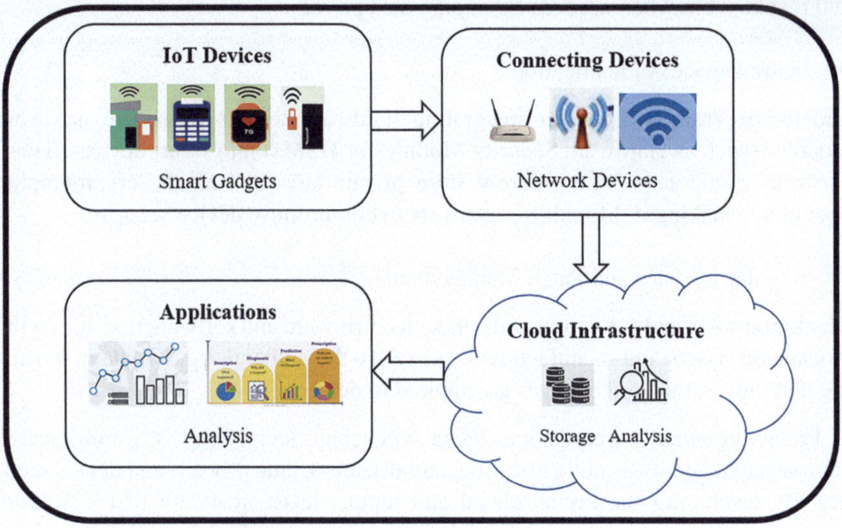

Fig. 1 IoT device with blockchain

4. Secure Device Updates

Firmware Verification Use blockchain to verify the authenticity and integrity of firmware updates. Smart devices can verify that the firmware update is signed with a private key associated with a trusted entity, preventing malicious updates.

5. Supply Chain Security

Blockchain Provenance Incorporate blockchain to track the provenance of smart devices throughout the supply chain. This ensures that devices are genuine and have not been tampered with during manufacturing or distribution.

6. Decentralization and Redundancy

Blockchain Nodes Distribute blockchain nodes across various smart home devices. This decentralization reduces the risk of a single point of failure and enhances the network's resilience against attacks.

7. Smart Contracts for Automation

Automation Rules Use blockchain-based smart contracts to create automation rules for smart devices. For example, a contract could specify that when a motion sensor detects movement, it triggers the lights to turn on. These contracts can be securely executed without the need for a centralized controller.

8. User Data Control

User Consent Blockchain can enable users to have fine-grained control over their data. They can grant temporary access permissions to certain devices or applications and revoke them when needed, enhancing data privacy.

9. Secure Device Authentication

Hardware Wallets Consider integrating hardware wallets or secure hardware modules (such as Hardware Security Modules or HSMs) into smart devices. These hardware components can securely store private keys and handle cryptographic operations, making it difficult for attackers to compromise device security.

10. Regular Updates and Patch Management

Blockchain Records Keep records of device firmware and software updates on the blockchain. Users and manufacturers can verify the authenticity of updates, ensuring that only authorized changes are applied to devices.

Protecting smart home devices using blockchain technology is a multifaceted approach that addresses authentication, authorization, data privacy, and device security. By leveraging the decentralized and tamper-resistant nature of blockchain, smart home ecosystems can be made more secure and trustworthy for users.

However, it is essential to ensure that the implementation is user-friendly and does not introduce unnecessary complexity for homeowners.

4.2 Enhancing Supply Chain Security

Case Study 2: Ensuring Authenticity in the Electronics Supply Chain
The electronics industry faces significant challenges related to supply chain security, including counterfeit components and unauthorized modifications. This case study examines how a multinational electronics manufacturer employed blockchain to address these issues.

Immutable Supply Chain Records Blockchain was used to record every stage of the supply chain, from component manufacturing to assembly. This transparency allowed stakeholders to verify the authenticity and integrity of components.

Anticounterfeiting Measures Blockchain-based product labeling and tracking enabled customers and distributors to verify the authenticity of purchased electronics.

Tamper Detection By integrating blockchain with hardware security modules, the manufacturer could detect and respond to any attempts at tampering with devices during transit.

The electronics industry faces significant challenges related to supply chain security, including counterfeit components, unauthorized modifications, and compromised integrity. Ensuring the authenticity of electronic components is crucial for maintaining product quality, safety, and reliability.

The primary objective of this case study is to demonstrate how a multinational electronics manufacturer implemented blockchain technology to address these supply chain challenges and enhance authenticity verification, as shown in the supply chain example in Fig. 2.

1. Immutable Supply Chain Records

Blockchain Integration The electronics manufacturer integrated blockchain technology into its supply chain management system. Each step of the supply chain, from component manufacturing to assembly, was recorded on the blockchain.

Transparency This blockchain-based transparency allowed all stakeholders, including manufacturers, distributors, and end-users, to access and verify the authenticity and integrity of components and products.

2. Anticounterfeiting Measures

Blockchain-Based Product Labeling Each electronic component was labeled with a unique identifier linked to a blockchain record. This identifier contained

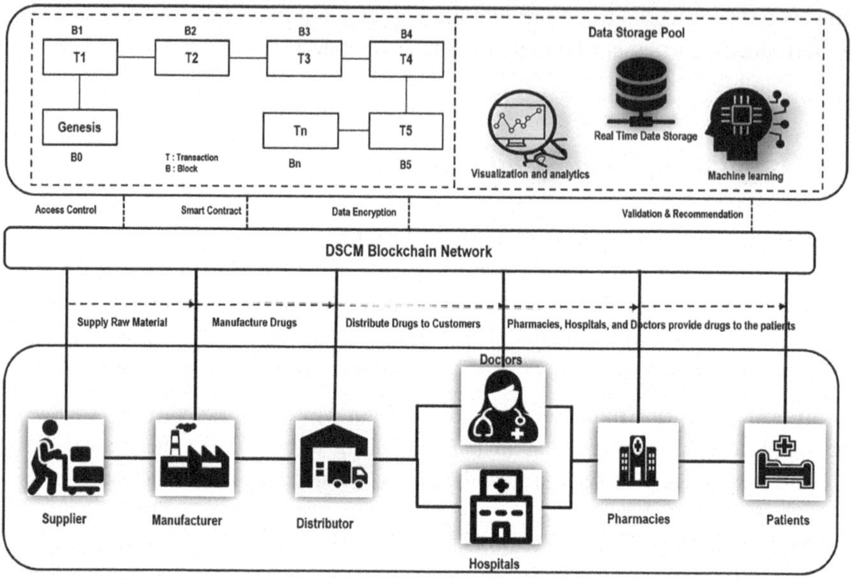

Fig. 2 Supply chain example with blockchain

information about the component's origin, manufacturing date, and other relevant data.

Verification by End-Users End-users could scan the product label using a smartphone app or website to verify the authenticity of the component. The app would check the blockchain for the component's history, ensuring it was genuine.

3. Tamper Detection

Integration with Hardware Security Modules (HSMs) To prevent physical tampering, HSMs were integrated into critical components. These HSMs would generate cryptographic signatures for the component's firmware and other data.

Blockchain Confirmation These signatures were recorded on the blockchain. Any attempts to tamper with the component would result in a mismatch between the recorded signature and the current state of the component, triggering alerts.

4. Vendor and Supplier Verification

Blockchain-Based Vendor Profiles The manufacturer maintained profiles for all its vendors and suppliers on the blockchain. These profiles included certification information and performance history.

Audits and Compliance Regular audits and compliance checks were conducted, and the results were recorded on the blockchain. This ensured that vendors adhered to quality and security standards.

5. Smart Contracts for Payments and Quality Assurance

Smart Contract Integration Smart contracts were employed to automate payments to suppliers based on predefined conditions. Payments were only released when quality and authenticity checks were successfully completed.

Quality Assurance Blockchain records tracked the quality assurance processes of critical components. If any component failed quality checks, the smart contract would trigger alerts and necessary actions.

Benefits

Counterfeit Prevention: The blockchain-based product labeling and verification system made it extremely difficult for counterfeit components to infiltrate the supply chain, ensuring product authenticity.

Enhanced Transparency: Transparency across the supply chain enhanced trust among stakeholders, and everyone had access to real-time data about component origins and quality.

Tamper Detection: The integration of HSMs and blockchain allowed for tamper detection at an unprecedented level, ensuring the integrity of components.

Efficient Payments and Quality Control: Smart contracts streamlined payment processes and quality assurance, reducing manual interventions and errors.

Regulatory Compliance: By maintaining auditable records on the blockchain, the manufacturer easily met regulatory compliance requirements.

This case study illustrates how blockchain technology can revolutionize supply chain management in the electronics industry. By ensuring the transparency, integrity, and authenticity of electronic components, the manufacturer not only enhanced product quality but also bolstered consumer trust and security, resulting in a more reliable and resilient supply chain ecosystem.

4.3 Smart Contracts for Hardware Security

Case Study 3: Secure Automated Transactions in the Automotive Industry

The automotive industry relies on complex supply chains, involving numerous stakeholders and transactions. This case study explores how a leading automotive manufacturer leveraged blockchain and smart contracts to streamline and secure its supply chain.

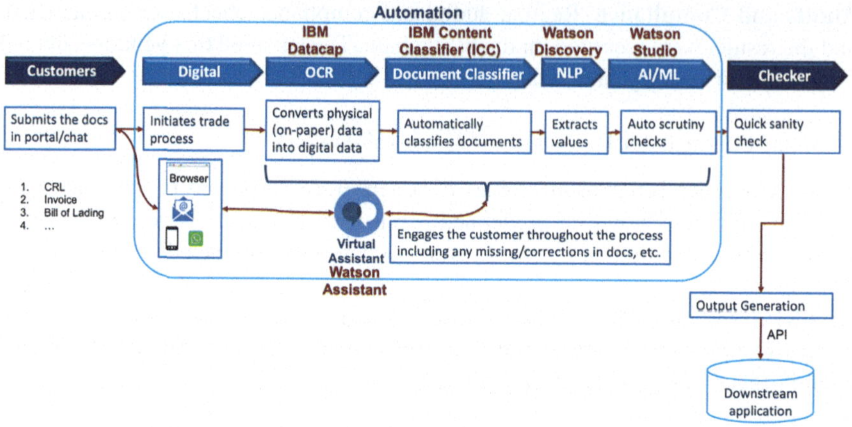

Fig. 3 Secure automated transactions in the automotive industry

Smart Contracts for Payments Smart contracts were used to automate payments to suppliers based on predefined conditions, reducing the risk of disputes and ensuring timely transactions.

Quality Assurance Blockchain-based records were used to track the quality assurance processes of critical components. Any deviations from predefined quality standards triggered alerts and actions through smart contracts.

Immutable Records for Regulatory Compliance Regulatory compliance requirements were met by using blockchain to create immutable records of all manufacturing and supply chain activities.

These case studies exemplify the versatility and transformative potential of blockchain technology in bolstering hardware security across diverse industries and use cases as shown in Fig. 3. By integrating blockchain into their hardware systems, organizations have not only mitigated security risks but also unlocked new levels of efficiency and transparency in their operations.

5 Privacy and Cryptographic Innovations

In the realm of hardware security, safeguarding privacy and data integrity is of paramount importance. This section explores the critical role of privacy-preserving protocols and cryptographic innovations, highlighting how they contribute to the enhanced security and trustworthiness of embedded systems integrated with blockchain technology.

5.1 Cryptography in Hardware Security

Cryptography is the cornerstone of modern security practices, and its role in hardware security cannot be overstated. By employing cryptographic techniques, embedded systems can protect sensitive data, ensure secure communications, and verify the authenticity of entities in a network.

Key aspects of cryptographic innovations in hardware security include the following:

Public Key Infrastructure (PKI): PKI provides a framework for secure communication and authentication. Hardware devices can utilize PKI to establish secure channels and verify the identity of peers.

Homomorphic Encryption: This advanced cryptographic technique allows computations to be performed on encrypted data without the need to decrypt it first. It offers privacy benefits in scenarios where data needs to be processed while remaining confidential.

Zero-Knowledge Proofs: Zero-knowledge proofs enable one party to prove to another party that they know a specific piece of information without revealing the information itself. This technology is valuable for identity verification without disclosing sensitive data.

Post-Quantum Cryptography: As quantum computing advances, post-quantum cryptography becomes essential for securing hardware systems against quantum threats. It involves cryptographic algorithms that are resistant to quantum attacks.

5.2 Privacy-Preserving Protocols

Privacy-preserving protocols play a pivotal role in ensuring that sensitive information remains confidential even in blockchain-based systems. These protocols enable secure interactions while hiding the actual data from prying eyes. Some notable privacy-preserving techniques include

Confidential Transactions This protocol conceals transaction amounts on the blockchain, making it difficult for external parties to deduce financial details.

Ring Signatures Ring signatures are used to obscure the sender's identity in a transaction. Multiple public keys are mixed, making it challenging to identify the true sender.

Zero-Knowledge Proof-Based Privacy Privacy coins and systems often utilize zero-knowledge proofs to verify transactions without revealing the transaction details. Examples include Zcash and Monero.

Secure Multiparty Computation (SMPC) SMPC protocols enable multiple parties to jointly compute a function over their inputs while keeping those inputs private.

These privacy-preserving protocols, when integrated into hardware systems, empower users with a higher level of control over their data and transactions, contributing to enhanced security and trust.

5.3 Blockchain's Role in Privacy

Blockchain, with its transparent and immutable ledger, may seem at odds with privacy. However, innovative approaches like confidential transactions, off-chain scaling solutions, and the use of privacy coins demonstrate that blockchain can indeed provide privacy features.

By integrating privacy features into the blockchain layer, hardware systems can leverage the advantages of blockchain while preserving the confidentiality of sensitive information. This is particularly relevant for industries and use cases where data privacy is non-negotiable, such as health care, finance, and identity management.

Incorporating privacy-preserving cryptography and blockchain privacy features into hardware systems enhances the overall security posture, ensuring that sensitive data remains confidential while benefiting from the trust and transparency offered by blockchain technology.

6 Mitigating Vulnerabilities

In the ever-evolving landscape of hardware security and blockchain technology, mitigating vulnerabilities is a critical priority. This section explores strategies and best practices for identifying, analyzing, and addressing vulnerabilities in hardware systems, ensuring their trustworthiness and resilience.

6.1 Identifying and Analyzing Hardware Vulnerabilities

Vulnerability identification is the first step in securing hardware systems. It involves assessing the system's components, configurations, and potential weaknesses. Common techniques and considerations include

Vulnerability Scanning Regular scans and assessments of hardware systems can reveal known vulnerabilities, misconfigurations, and outdated software or firmware.

Threat Modeling Developing threat models specific to the hardware system helps identify potential attack vectors and prioritize security measures.

Penetration Testing Ethical hacking and penetration testing simulate real-world attacks to identify vulnerabilities and weaknesses in hardware systems.

Code and Design Reviews Reviewing the source code, firmware, and hardware designs can uncover vulnerabilities, particularly those related to implementation flaws.

6.2 *Strategies for Vulnerability Mitigation*

Once vulnerabilities are identified, effective mitigation strategies are crucial to reduce risks and enhance the security of hardware systems integrated with blockchain technology. Key strategies include

Patch Management Implement a robust patch management process to ensure timely updates and fixes for known vulnerabilities. This includes both hardware and software/firmware components.

Secure Boot and Trusted Execution Environments (TEEs) Secure boot processes, combined with Trusted Execution Environments (such as Intel SGX or ARM TrustZone), establish a trusted foundation for hardware systems by verifying the integrity of firmware and software during boot-up.

Least Privilege Principle Enforce the principle of least privilege to restrict access and permissions to the minimum necessary for each user or component. This limits potential damage in case of a compromise.

Zero Trust Architecture Adopt a Zero Trust approach, where trust is never assumed, and verification is required from anyone trying to access resources within the hardware system.

Blockchain Auditing Implement auditing and monitoring mechanisms that continuously assess the integrity and activity of blockchain transactions and smart contracts, allowing for the detection of suspicious behavior.

Incident Response Plans Develop comprehensive incident response plans to address vulnerabilities and breaches promptly. These plans should include steps for containment, eradication, and recovery.

Security Training Train personnel involved in managing and maintaining hardware systems on security best practices and how to recognize and respond to security incidents.

Secure Supply Chain Implement rigorous supply chain security measures to ensure the integrity of hardware components from manufacturing to deployment. This includes measures to detect and prevent counterfeit or compromised components.

6.3 Ensuring Trustworthy Hardware Systems

Ultimately, the goal of vulnerability mitigation is to ensure the trustworthiness of hardware systems. Trustworthy systems are those that can resist attacks, maintain data confidentiality, and provide a high level of integrity. Strategies for achieving trustworthy hardware systems include

Continuous Monitoring Implement continuous monitoring solutions that can detect and respond to anomalies and security incidents in real time.

Secure Development Practices Integrate security into the entire development life cycle of hardware systems, from design and coding to testing and deployment.

Security by Design Embed security considerations into the architecture and design of hardware systems from the outset, rather than attempting to retrofit security later.

Compliance and Standards Adhere to industry standards and compliance requirements relevant to hardware security and blockchain, ensuring that systems meet recognized security benchmarks.

Regular Updates and Maintenance Maintain hardware systems with regular updates, patches, and firmware upgrades to address emerging vulnerabilities and threats.

By implementing these strategies and best practices, organizations can significantly reduce the risk of vulnerabilities and enhance the overall security and trustworthiness of their hardware systems integrated with blockchain technology.

7 Future Dircctions

As the digital landscape continues to evolve, the fusion of blockchain technology and hardware security promises to shape the future of cybersecurity. In this section, we explore the emerging trends, innovative advancements, and potential developments that will define the path forward for blockchain in hardware security.

7.1 *Evolving Landscape of Blockchain and Hardware Security*

The intersection of blockchain and hardware security is a dynamic field, continually influenced by technological advancements, evolving threats, and regulatory changes. Key trends and directions include

Quantum-Resistant Cryptography With the advancement of quantum computing, the need for quantum-resistant cryptography will become increasingly urgent. Research and development in this area are expected to intensify.

Integration with 5G and Edge Computing The proliferation of 5G networks and edge computing will create new opportunities and challenges for hardware security. Blockchain may play a pivotal role in securing decentralized edge environments.

Interoperability and Standardization The development of interoperability standards and protocols for blockchain integration into hardware systems will gain momentum, enabling seamless and secure cross-platform interactions.

Decentralized Identity and Self-Sovereign Identity (SSI) Blockchain-based solutions for decentralized identity management and SSI are poised to revolutionize how individuals and organizations control and protect their digital identities.

Regulatory Frameworks Governments and regulatory bodies will continue to develop and refine frameworks for blockchain technology, affecting how it is employed in hardware security and compliance.

7.2 *Future Applications and Innovations*

The future of blockchain in hardware security holds promise of groundbreaking applications and innovations:

Blockchain-Based Hardware Security as a Service (HSaaS): HSaaS platforms may emerge, offering scalable and customizable hardware security solutions powered by blockchain technology.

AI and Blockchain Synergy: The integration of artificial intelligence (AI) and blockchain will enable predictive and adaptive security measures, enhancing threat detection and response.

Blockchain in Edge Devices: The deployment of blockchain in edge devices and IoT sensors will enable decentralized, autonomous, and secure decision-making at the edge of networks.

Cross-Industry Collaboration: Collaboration between industries, such as healthcare, finance, and supply chain, will lead to cross-industry blockchain solutions for comprehensive hardware security.

7.3 Ethical and Regulatory Considerations

The ethical and regulatory landscape surrounding blockchain and hardware security will continue to evolve:

Privacy Regulations: As privacy concerns intensify, regulations governing data protection and user consent will become more stringent, influencing how blockchain systems handle personal information.

Ethical Use of Data: Ethical considerations will drive discussions on the responsible use of blockchain-enabled hardware systems, particularly in areas like health care and biotechnology.

Global Harmonization: Efforts to harmonize international regulations and standards for blockchain technology will gain importance to facilitate cross-border operations.

In conclusion, the future of blockchain in hardware security is a dynamic landscape characterized by innovation, evolving threats, and ethical considerations. Staying at the forefront of these developments and embracing emerging technologies will be essential for organizations seeking to fortify their hardware systems against the evolving challenges of the digital world.

8 Conclusion

In the realm of hardware security, where the stakes are high and vulnerabilities can have profound consequences, the convergence of blockchain technology has ushered in a new era of trust and resilience. This chapter has navigated the intricate landscape where these two domains intersect, uncovering the transformative potential and real-world applications that redefine how we secure our digital infrastructure. From the fundamental principles of hardware security to the integration of blockchain into embedded systems, we have explored the building blocks of trustworthiness in hardware. We have witnessed how blockchain technology, with its decentralization, immutability, and consensus mechanisms, offers a robust foundation upon which hardware security can thrive. The case studies presented in this chapter have underscored the versatility of this fusion. Whether it is securing IoT devices in our homes, fortifying the integrity of the electronics supply chain, or automating secure transactions in the automotive industry, blockchain's impact on hardware security is tangible and transformative. Privacy-preserving protocols and cryptographic innovations have illuminated the path to confidentiality and data integrity in an increasingly connected world. These innovations ensure that sensitive information remains confidential even in blockchain-based systems, fostering a new era of privacy and security. Mitigating vulnerabilities, a perpetual challenge in the ever-evolving landscape of hardware security, is addressed through comprehensive strategies and best practices. From vulnerability identification to incident response, the blueprint for secure hardware systems has been laid out, ensuring the trustworthiness of these systems. As we peer into the future, we see emerging trends that promise to shape the landscape further. Quantum-resistant cryptography, the integration of blockchain with 5G and edge computing, and the evolution of regulatory frameworks all underscore the dynamic nature of this field. The possibilities are boundless, from hardware security as a service to the synergy of AI and blockchain, and the decentralized decision-making capabilities of edge devices. Yet, we must navigate this transformative journey with ethical considerations at the forefront, ensuring the responsible and ethical use of data in an increasingly interconnected world. In this ever-changing landscape, one thing remains constant: the importance of securing our hardware systems. As we conclude this chapter, we are reminded that the journey to hardware security and trust is an ongoing one, filled with challenges and opportunities. It is a journey where innovation, adaptability, and a commitment to the highest ethical standards will be our guiding stars. The fusion of blockchain and hardware security represents not only a technological breakthrough but also a testament to human ingenuity in the face of evolving threats. With each advancement, we inch closer to a digital world where trust is not an aspiration but an inherent attribute of our hardware systems. This chapter serves as a compass for that journey, illuminating the path forward for all those dedicated to securing the foundation of our digital future.

References

1. Kulkarni, A., Hazari, N. A., & Niamat, M. (2019, July). A Blockchain Technology Approach for the Security and Trust of the IC Supply Chain. In *2019 IEEE National Aerospace and Electronics Conference (NAECON)* (pp. 249–252). IEEE.
2. Cabrera-Gutiérrez, A. J., Castillo, E., Escobar-Molero, A., Álvarez-Bermejo, J. A., Morales, D. P., & Parrilla, L. (2022). Integration of hardware security modules and permissioned blockchain in industrial IoT networks. *IEEE Access, 10*, 114331–114345.
3. Singh, J., & Michels, J. D. (2018, April). Blockchain as a service (BaaS): Providers and trust. In *2018 IEEE European Symposium on Security and Privacy Workshops (EuroS&PW)* (pp. 67–74). IEEE.

4. Alrubei, S., Ball, E., & Rigelsford, J. (2022, November). Adding Hardware Security into IoT-Blockchain Platforms. In *2022 IEEE Latin-American Conference on Communications (LATINCOM)* (pp. 1–6). IEEE.
5. Xu, M., Guo, Y., Hu, Q., Xiong, Z., Yu, D., & Cheng, X. (2023). A trustless architecture of blockchain-enabled metaverse. *High-Confidence Computing, 3*(1), 100088.
6. Saha, S. S., Gorog, C., Moser, A., Scaglione, A., & Johnson, N. G. (2021, April). Integrating hardware security into a blockchain-based transactive energy platform. In *2020 52nd North American Power Symposium (NAPS)* (pp. 1–6). IEEE.
7. Sklavos, N., Chaves, R., Di Natale, G., & Regazzoni, F. (2017). *Hardware security and trust.* Springer.
8. Jesus, V. (2018, November). Blockchain-enhanced roots-of-trust. In *2018 International Conference on Smart Communications and Networking (SmartNets)* (pp. 1–7). IEEE.
9. Abed, S. E., Jaffal, R., & Mohd, B. J. (2023). A review on blockchain and iot integration from energy, security and hardware perspectives. *Wireless Personal Communications, 129*(3), 2079–2122.
10. Islam, M. N., & Kundu, S. (2020). IoT security, privacy and trust in home-sharing economy via blockchain. In K. K. Choo, A. Dehghantanha, & R. Parizi (Eds.), *Blockchain Cybersecurity, Trust and Privacy* (pp. 33–50). Springer.
11. Enkhtaivan, B., Takenouchi, T., & Sako, K. (2019, August). A fair anonymous auction scheme utilizing trusted hardware and blockchain. In *2019 17th International Conference on Privacy, Security and Trust (PST)* (pp. 1–5). IEEE.
12. Bentov, I., Ji, Y., Zhang, F., Breidenbach, L., Daian, P., & Juels, A. (2019, November). Tesseract: Real-time cryptocurrency exchange using trusted hardware. In *Proceedings of the 2019 ACM SIGSAC Conference on Computer and Communications Security* (pp. 1521–1538).
13. Gountia, D. (2019). Towards scalability trade-off and security issues in state-of-the-art blockchain. *EAI endorsed Transactions on Security and Safety, 5*(18), e4–e4.
14. Cui, P., Guin, U., Skjellum, A., & Umphress, D. (2019). Blockchain in IoT: current trends, challenges, and future roadmap. *Journal of Hardware and Systems Security, 3*, 338–364.
15. Boireau, O. (2018). Securing the blockchain against hackers. *Network Security, 2018*(1), 8–11.
16. Novotny, P., Zhang, Q., Hull, R., Baset, S., Laredo, J., Vaculin, R., et al. (2018). Permissioned blockchain technologies for academic publishing. *Information Services & Use, 38*(3), 159–171.
17. Farah, M. B., Ahmed, Y., Mahmoud, H., Shah, S. A., Al-Kadri, M. O., Taramonli, S., et al. (2024). *A survey on blockchain technology in the maritime industry: Challenges and future perspectives*. Future Generation Computer Systems.
18. Dhar Dwivedi, A., Singh, R., Kaushik, K., Rao Mukkamala, R., & Alnumay, W. S. (2024). Blockchain and artificial intelligence for 5G-enabled Internet of Things: Challenges, opportunities, and solutions. *Transactions on Emerging Telecommunications Technologies, 35*(4), e4329.
19. Nuriddinov, A. (2024). Blockchain Applications in the sports industry. *PEDAGOG, 7*(5), 21–30.
20. Kasula, B. Y., Whig, P., Vegesna, V. V., & Yathiraju, N. (2024). unleashing exponential intelligence: Transforming businesses through advanced technologies. *International Journal of Sustainable Development Through AI, ML and IoT, 3*(1), 1–18.
21. Bennet, D., Maria, L., Sanjaya, Y. P. A., & Zahra, A. R. A. (2024). Blockchain technology: Revolutionizing transactions in the digital age. *ADI Journal on Recent Innovation, 5*(2), 192–199.
22. Xu, M., Guo, Y., Liu, C., Hu, Q., Yu, D., Xiong, Z., et al. (2024). Exploring blockchain technology through a modular lens: A survey. *ACM Computing Surveys, 56*(9), 1–39.
23. Wang, X., Wu, Y. C., & Ma, Z. (2024). Blockchain in the courtroom: Exploring its evidentiary significance and procedural implications in US judicial processes. *Frontiers in Blockchain, 7*, 1306058.
24. Whig, P., Battina, D. S., Venkata, S., Bhatia, A. B., & Alkali, Y. J. (2024). Role of intelligent IoT applications in fog computing. *Fog Computing for Intelligent Cloud IoT Systems, 99.*

25. Zhu, L., Feng, T., Lu, Y., & Jiang, W. (2024). Using blockchain or not? A focal firm's blockchain strategy in the context of carbon emission reduction technology innovation. *Business Strategy and the Environment, 33*(4), 3505–3531.
26. Ani, N., Millah, S., & Sunarya, P. A. (2024). Optimizing Online Business Security with Blockchain Technology. *Startupreneur Business Digital (SABDA Journal), 3*(1), 67–80.
27. Rasel, M., & Bommu, R. (2024). Blockchain-Enabled Secure Interoperability: Advancing Electronic Health Records (EHR) Data Exchange. *International Journal of Advanced Engineering Technologies and Innovations, 1*(3), 262–281.
28. Babu, R. M., Satamraju, K. P., Gangothri, B. N., Malarkodi, B., & Suresh, C. V. (2024). A hybrid model using genetic algorithm for energy optimization in heterogeneous internet of blockchain things. *Telecommunications and Radio Engineering, 83*, 1–16.
29. Whig, P., Velu, A., Nadikattu, R. R., & Alkali, Y. J. (2024). Role of AI and IoT in Intelligent Transportation. In *Artificial Intelligence for Future Intelligent Transportation* (pp. 199–220). Apple Academic Press.
30. Whig, P., Gera, R., Bhatia, A. B., & Reddy, R. (2024). Convergence of Blockchain and IoT in Healthcare. In *Convergence of Blockchain and Internet of Things in Healthcare* (p. 277). CRC Press.
31. Kayikci, S., & Khoshgoftaar, T. M. (2024). Blockchain meets machine learning: a survey. *Journal of Big Data, 11*(1), 9.
32. Wahrstätter, A., Ernstberger, J., Yaish, A., Zhou, L., Qin, K., Tsuchiya, T., et al. (2024, May). Blockchain censorship. In *Proceedings of the ACM on Web Conference 2024* (pp. 1632–1643).
33. Whig, P., Velu, A. R. U. N., & Sharma, P. (2022). Demystifying federated learning for blockchain: A case study. In *Demystifying federated learning for Blockchain and industrial internet of things* (pp. 143–165). IGI Global.
34. Whig, P., Velu, A., & Nadikattu, R. R. (2022). Blockchain platform to resolve security issues in IoT and smart networks. In *AI-enabled agile internet of things for sustainable FinTech ecosystems* (pp. 46–65). IGI Global.
35. Gadekallu, T. R., Huynh-The, T., Wang, W., Yenduri, G., Ranaweera, P., Pham, Q. V., ... & Liyanage, M. (2022). Blockchain for the metaverse: A review. arXiv preprint arXiv:2203.09738.
36. Guo, H., & Yu, X. (2022). A survey on blockchain technology and its security. *Blockchain: Research and Applications, 3*(2), 100067.
37. Gad, A. G., Mosa, D. T., Abualigah, L., & Abohany, A. A. (2022). Emerging trends in blockchain technology and applications: A review and outlook. *Journal of King Saud University-Computer and Information Sciences, 34*(9), 6719–6742.
38. Gorkhali, A., Li, L., & Shrestha, A. (2020). Blockchain: A literature review. *Journal of Management Analytics, 7*(3), 321–343.
39. Pawan, W. (2023). Peer-to-peer energy trading with blockchain: a case study. In *Blockchain-Based Systems for the Modern Energy Grid* (pp. 171–188). Academic.
40. König, L., Unger, S., Kieseberg, P., Tjoa, S., & Blockchains, J. R. C. (2020). The Risks of the Blockchain A Review on Current Vulnerabilities and Attacks. *Journal of Internet Services and Information Security, 10*(3), 110–127.

Hardware-Assisted Security of Smart IoT Applications

Amit Singh

1 Introduction

Earth, the planet is heading toward future and at a hypercritical stage nowadays, where everyone is in some way hurting/damaging our society and the digital world is no more secure too. Safeguarding of encircled things, shown in Fig. 1, is the prime concern of today's world. Presumptions and expectations are some of the thinking behaviors that lead our society in the direction of disaster. In this challenging living world, 360° security is one of the biggest and unavoidable challenges. If the security of digital and electronic solution space available around us is compromised, it can lead to sufficient harm to an individual or the society as a whole.

Fig. 1 shows how much an individual is enclosed with the surrounding particulars that are prone to be compromised, and if so, the consequences can be eventually thought of. From online banking and transactions to automated home and offices, smart transportation to online shopping, etc. is associated with Internet and so the World Wide Web.

The security issues can be categorized as hardware security and software security issues, and based on various solutions, they are proposed and employed at different levels.

As IoT devices are more prone to compromise with unethical attacks, hardware security provides various benefits as mentioned below [1]:

(i) Strong and impenetrable secret key storage in physical devices.
(ii) Cryptographic functions integrated with the devices.
(iii) Safeguarding using suitable, modern screening measures against physical threats, such as power analysis or tampering.

A. Singh (✉)
Department of CSE, School of Technology, Pandit Deendayal Energy University, Raysan, Gujarat, India

© The Author(s), under exclusive license to Springer Nature Switzerland AG 2025

A. Mishra et al. (eds.), *Hardware Security: Challenges and Solutions*,
https://doi.org/10.1007/978-3-031-81213-2_3

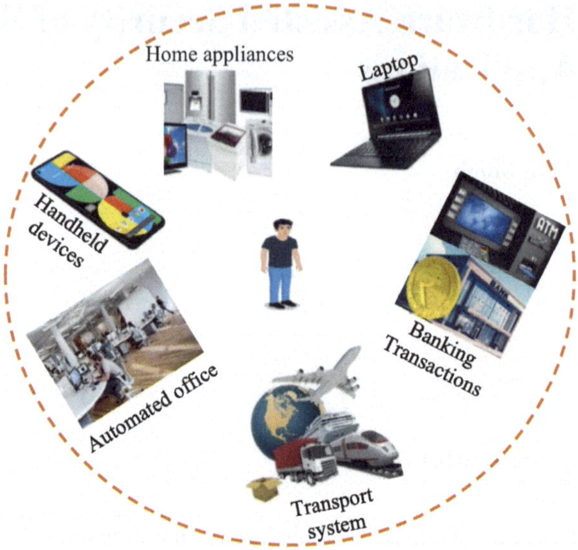

Fig. 1 Dimensions of security threat to individuals

(iv) Support for device life cycle management that is permitted and authenticated, including downloading, activating, modifying, and canceling contracts and memberships. Moreover, the device must ensure:

(a) ***Authenticity:*** assurance that the sensor or equipment used to transmit the message is authentic and well-known.
(b) ***Integrity:*** assurance of unaltered message is received.
(c) ***Confidentiality:*** assurance that the message is not intercepted and viewed by anyone else during communication.

Some prominent problems and solution approaches in hardware-assisted security are explored and briefly highlighted in the following sections. Section 2 is dedicated to the fundamental concept of IoT and smart things. On top of software security mechanisms, hardware-assisted security approaches in various IoT application domains, such as health care, autonomous vehicles, and smart grid are discussed in Sect. 3. Some trade-off parameters with hardware-assisted security mechanisms are highlighted in Sect. 4. Finally, Section 5 provides bullet points of the chapter and suggests various possible research directions in the field of hardware-assisted IoT security.

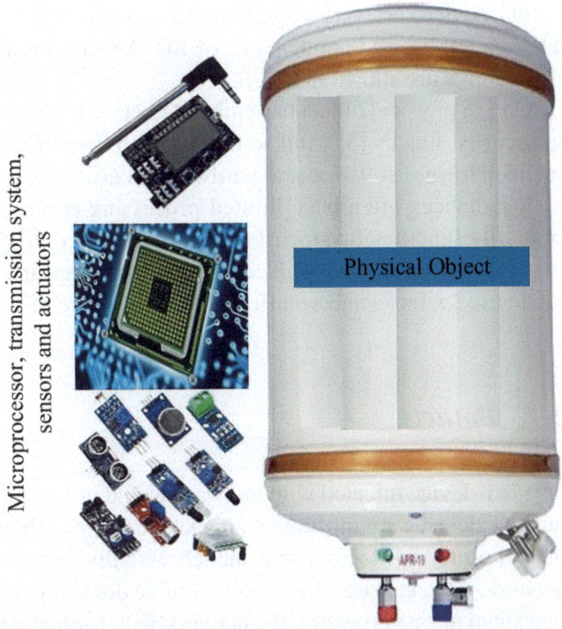

Fig. 2 Electronic devices employed in an electric geyser to make a smart geyser

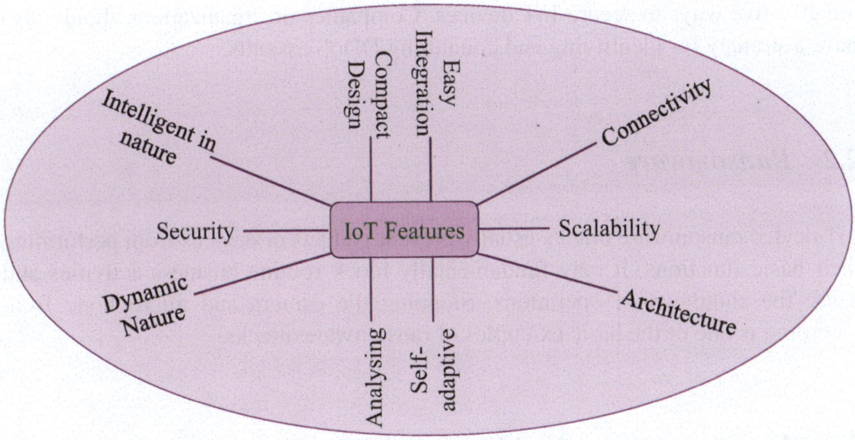

Fig. 3 IoT features

2 Internet of Things and Security Attacks

Internet of Things (IoT) is defined as the interconnection of physical objects so that they can communicate through message passing [2]. In an IoT environment, a physical object has associated electronic sensors, actuators and microcontroller devices that make objects connected, as shown in Fig. 2.

IoT devices are very popular nowadays due to its smart and intelligent features. Though the list and applications of IoT devices are not limited, however, a few of the features are shown in Fig. 3.

Once a physical object has embedded electrics and components, it is easily prone to security attacks [3]. And so the research community and industry are nowadays inclined toward software and hardware security solutions.

IoT devices often have limited processing power and storage capacity and are primarily functionality-focused machines. Hence, IoT devices are more vulnerable to security attacks than high computing machines and computers. Here are the following IoT attacks and security risks to be aware of.

2.1 Botnets

One IoT device infected with malware would not do much good; however, if numerous IoT devices are infiltrated, attackers can create botnets, which become an army of compromised devices that launch assaults on other systems connected to the network. Attackers can direct their zombie devices to carry out attacks like transmitting spam or overcrowding the network with traffic by using botnets to do the heavy lifting. Updating firmware, limiting device access, and changing default passwords are effective ways to secure IoT devices. Companies or organizations should also have a strategy for identifying and countering DDoS assaults.

2.2 Ransomware

IoT device ransomware threats usually prevent gadgets or devices from performing their basic functions. It may fundamentally block routine business activities and cause the shutdown of operations. Stopping the camera and microphone from recording is one of the basic examples of ransomware attacks.

2.3 Inadequate Security Measures in Devices and Unencrypted Data

The local storage of data in IoT devices is very rare, and in the absence of strong encryption mechanisms, the traffic is especially susceptible to interceptions, espionage, and hijacking. The best course of action is to enclose these kinds of devices with a customized shield that serves as a firewall, network address translator, and authentication mechanism to provide extra security against network discovery.

2.4 Rogue and Legacy Issues

IoT devices may start behaving dishonestly during their deployment. It may leave the dedicated group at any time and start performing some dangerous work for someone else. The attacks on IoT devices can be categorized as follows:

Physical Attacks: These types of attacks happen over the physical part of IoT devices. Node tempering and jamming, social engineering, sleep deprivation, and RF interferences largely come into this category.

Network Attacks: These types of attacks encroach network to steal or modify data. RFID spoofing, cloning, unauthorized access, DoS, and traffic analysis attacks come into this category.

Software Attacks: Such attacks contain a piece of computer code or applications that can damage IoT devices or steal information. Trojans, viruses, worms, spyware, and malicious scripts come into this category of attacks.

Encryption Attacks: It is a technique for getting beyond a cryptographic system's security by identifying a flaw in a key management system, cipher, code, or cryptographic protocol. Man-in-middle, cryptanalysis, and side channels come in this category of attacks.

Initially, software solutions has been tried to address security threats, however, due to more advanced software attacks these solutions become fade [4]. Hence, the hardware solutions came into the context to better address such threats and risks.

For a deeper insight into hardware solutions, let us first have a look over the problems, i.e., state-of-the-art IoT security threats:

Code Reuse Attacks: It uses code already present in the device to deliver a malicious payload.

Firmware Modification: In contrast to code reuse attacks, firmware modification does not require the gadget catalog. An attacker can run any arbitrary and malicious code with any limitation.

Brute Force Attack: Security can be breached by attempting numerous times and is very uncommon in practice due to computationally longer complexity ($O(n)$).

The command flow or control flow of the executing program in memory can be altered by the malicious code injection by the attackers [5]. Such kind of attacks control the device and force it to skip some instructions to manipulate the working of the physical device. Some authorization mechanisms are required to overcome such issues if they occur. The following section explores such research work from the existing literature.

3 Hardware-enabled Authorization in IoT

Hardware authentication is a method of user verification and validation that grants access to computer resources using a simple password and a specific physical item (like a token) owned by an authorized user. Some of the hardware-assisted security techniques are discussed as follows:

Physical Unclonable Functions [6–8]: PUF is a physical entity whose operation cannot be physically duplicated or cloned. It offers a physically defined "digital fingerprint" as an output (response) for a specific set of input data and conditions (challenge). It acts as a distinctive identification, usually for semiconductor devices like microprocessors. PUFs are frequently based on distinct physical dissimilarities that happen during the fabrication of semiconductors. A PUF is contained within a material structure. PUFs included in integrated circuits can be applied to strengthen security needs. These applications include privacy protection and cryptography in IoT devices.

PUFs are a novel class of low-cost hardware security primitives that are relatively suitable for smart and energy-constrained IoT devices due to their simple layout design. The deployment of PUFs is explored in IoT devices for authentication and message encryption purposes [9–11].

The security, functionality, and dependability of IoT-based consumer electronic products, as well as their numerous applications, are now gravely challenged by newly discovered threats, as shown in Fig. 4.

For device authentication and secret key generation, silicon PUFs are one-way physical functions that tie an input challenge (c_i) to an output response (r_i) by considering variations in the manufacturing process used to produce different semiconductor integrated circuits (ICs). Another variation in PUFs is presented by Maiti et al. [12] in the form of the Ring Oscillator (RO) PUF which is widely accepted by the research community. ROPUF can be readily implemented on application-specific integrated circuits (ASICs) and field-programmable gate arrays (FPGAs) due to its high performance with simple design requirements, isolating these devices from assaults of this nature.

PUF-based IoT hardware security challenges are somehow addressed; however, security certainty in IoT applications could be adversely affected by unreliability problems with the secret keys produced by the suggested PUF. Modeling attacks, cloning of challenges, and response pairs are some of the most common attacks discovered in the PUF design [13]. These drawbacks are addressed by Amsaad et al. [14, 15] and Maiti et al. [16]. Authors presented a lightweight Configurable ROPUFs (CF-ROPUFs) that utilize dedicated Field Programmable Gate Array (FPGA) resources for area-critical and trusted designs to protect IoT devices. They suggested many challenge–response pairs to improve the security and trust of IoT devices. The proposed framework is experimented with a 1-out-of-r coding algorithm over 30 Xilinx Spartan-3E FPGA chips, demonstrating its validity. The

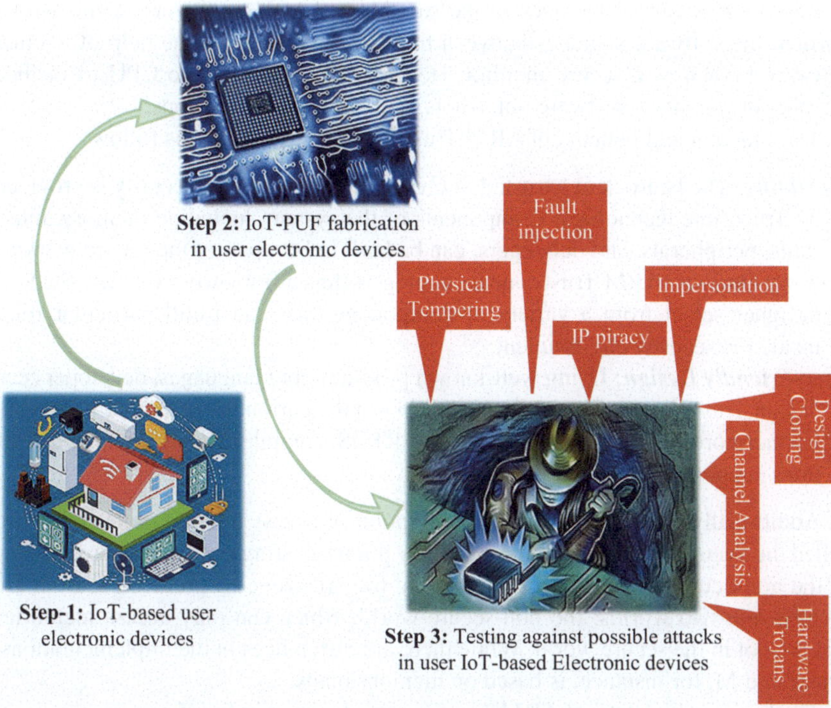

Fig. 4 Embedding and security tolerance demonstration of PUF-IoT

findings show a magnified average reproducibility of 99.63% and a high average repeatability of 98.2%.

A low-cost dynamic permutation approach for IoT devices to concurrently defy side-channel analysis and hardware Trojan attacks has been made [17]. The suggested approach strengthens the IoT processing unit against hardware attacks and obtains 5.8X higher cumulative partial guessing entropy.

Several solutions investigate hardware-based techniques for secure offloading, ML-based IoT authentication, access control, and malware detection approaches to safeguard an IoT infrastructure [18]. The authors highlighted the benefits of various strategies to address IoT security in a more adaptable and efficient manner.

ARM TrustZone [19] Modern embedded systems-on-a-chip (SoCs) are designed with compliance with the ARM Trusted Base System architecture. These criteria include base SoC isolation, cryptographic trusted boot, and debug protection. This helps against software and hardware risks. The ARM TrustZone hardware organization divides internal and external memories into trusted and nontrusted worlds, allowing the target application to be isolated from other applications during runtime.

ARM TrustZone technology is widely employed in a huge amount of application processors. It is used to secure program code and sensitive data. Software on ARM

Cortex-A processor either lives in the safe/secure world or the nonsecure world. Further, the software switches between these two worlds with the help of another software known as a secure monitor. Hence, it goes beyond the CPU to include peripherals, memory, software, interrupts, and SoC bus transactions.

The fundamental benefits of ARM TrustZone technology are as follows:

Flexibility: The bedrock of a trusted SoC ensuring system-wide security is provided by TrustZone technology. Components of the system, including memory, interrupts, peripherals, and debuggers, can be built to be a part of the secure world.

Security Options: ARM TrustZone technology offers a framework that lets the SoC designer select from a variety of components that can fulfill particular roles inside the security environment.

User-Friendly Design: Using well-known programming languages, developers can design TrustZone systems, while adhering to the current programmer paradigm. Furthermore, an extensive ecosystem of RTOS, compiler, debug, and trace solutions supports TrustZone.

Additionally, Trustzone-A has a single point of access into the protected world called the Security Monitor. Numerous entry points existing in a specific area known as the nonsecure callable region is possible for Trustzone-M processors. Both are divided into two worlds: the non-secure world, which can only access memories that are not in the secure world. While there are differences in the implementations, TrustZone-M, for instance, is based on memory maps.

The basic architecture of ARM TrustZone is depicted in Fig. 5.

An extended version of ARMv8-M [20] includes a Secure Attribute Unit (SAU) and an optional unit, i.e., Implementation-Defined Attribute Unit (IDAU). The address space of the ARMv8-M-based microcontroller is flagged as secure, nonsecure, and secure and nonsecure callable modes. Initially, when the software starts up, it begins in secure mode and further sets up SAU and IDAU if present. In such a way, it provides hardware-backed isolation between the program code and data. ARMv8-M uses secure gate (sg) instructions to change the mode of the CPU and

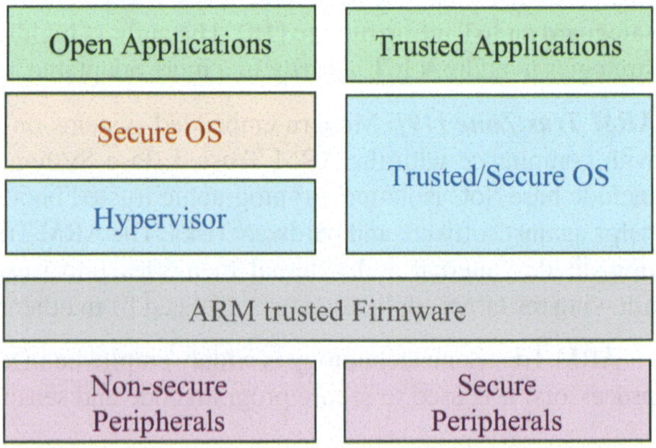

Fig. 5 Architecture of ARMv8-M

the transition state. The sg instruction provides fundamental defense against the accidental use of secure software.

Trusted Platform Module [21] Personal Computer (PC) security is aided by Trusted Platform Module (TPM) technology, which provides hardware-level defense against malware and advanced cyberattacks. Modern CPUs can incorporate TPM technology, which "securely stores artifacts used to authenticate the platform." TPMs provide various artifacts, including fingerprints, certificates, passwords, and all vital information that users wish to safeguard.

Though TPM technology has been used in enterprise IT for a long time, however, Microsoft feels a requirement for everyone, including small- and medium-sized business classes. New hardware-based security requirements that increase PC security are now available to Windows 11 users. A hardware-based security requirement for all Windows 11 PCs is the need for TPM 2.0 for the operating system to function.

A trusted platform module, also known as a microcontroller or a physical security device, is a component deployed in the motherboard or processor of a computer. To enable platform authentication, TPMs utilize cryptography to secure the critical data storage on PCs. To protect it from outside threats, they store a range of sensitive data behind a physical barrier, including user credentials, passwords, fingerprints, certificates, encryption keys, and other crucial work.

A set of specifications known as the Trusted Platform Module (TPM) [22] defines the operational principles of a stand-alone integrated circuit that offers secure storage and cryptography services to the applications. TPM offers services to create random numbers, safe cryptographic key generation, remote attestation, and key binding and sealing. TPM routines can access the registers and modify them through software patches. The hashing feature is exploited by the TPM routines to ensure security. If the software is in an unknown execution state, the TPM routines will not uncover secrecy.

Fig. 6 presents a broader view of the TPM module.

For integrity protection, attestation has also been suggested in conjunction with hardware security managers (HSMs) and physically unclonable functions (PUFs) [23]. HSM units are used to coordinate not only the distribution of keys but also cryptography-related processes including trusted boot, authentication, and debug protection.

Cyber-physical systems (CPSs) connect physical infrastructure and things to the Internet and one another by integrating sensing, computation, control, and networking [24, 25]. CPS-enabled IoT devices are attached to smart grids, industries, and automobiles.

Though artificial intelligence (AI)-, machine learning (ML)-, and deep learning (DL)-based solutions can increase overhead over IoT devices in terms of computing as well as energy resources, some of the solutions are proposed in the literature [26–31]. However, there are several obstacles in employing machine learning techniques to secure Internet of Things environments. These arise from the nuanced characteristics of different types of assaults as well as the complexity of hardware and software circuitry that has security-weak spots.

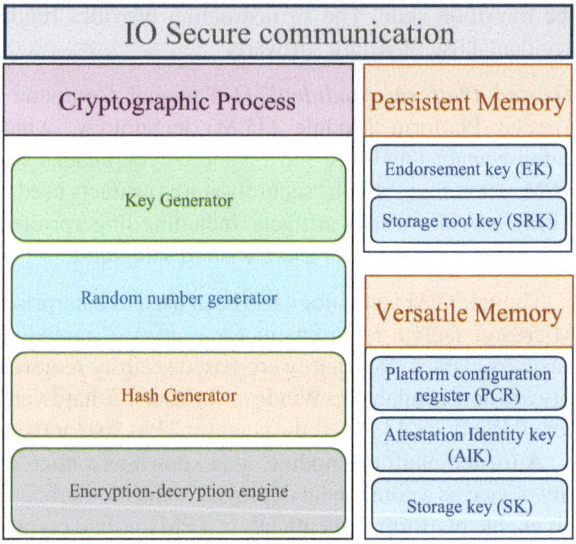

Fig. 6 Components of Trusted Platform Module (TPM)

IoT infrastructures are growing rapidly in the automotive, health care, and industrial sectors as a result of the integration of many networks and devices. This makes it practically impossible to discover aberrant data that could result in bodily harm due to unidentified attack vectors and attack strategies [32–34].

Similarly, IoT security issues and hardware-assisted solutions are discussed in a guest editorial by Sengupta et al. [35]. Dong et al. [36] proposed a multilayer hardware Trojan protection framework for IoT chip devices.

Security is one of the biggest challenges in IoT and needs to be addressed. Software security can alone not be sufficient to protect user data and communication, and hence, the hardware-assisted devices are required on top of the software solutions. Some of the IoT-based sectors with hardware-assisted security systems are categorically presented in the next section.

3.1 Smart Healthcare System

Through the integration of diverse technologies such as blockchain, big data, and machine learning (ML), the implementation of IoMT has advanced healthcare systems. Figure 7 shows the various components of the smart healthcare system.

The healthcare monitoring system critically demands a preventive approach against the following:

(i) The possible cyberattack on the sensors implanted within or surrounding the human body.
(ii) Sophisticated and lightweight cryptographic mechanism for the sensitive data transmission between the patient and doctor.

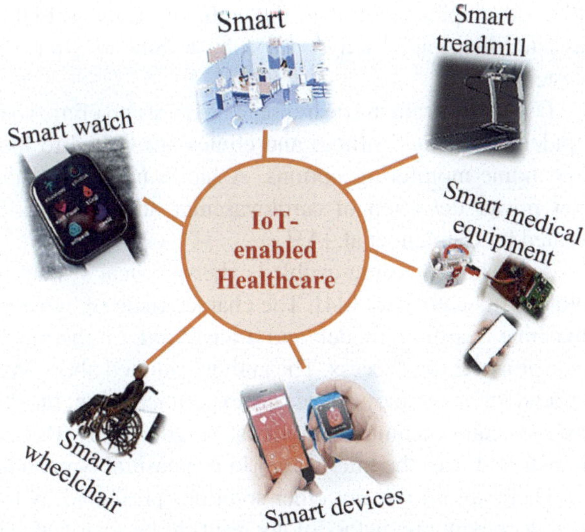

Fig. 7 IoT support in smart healthcare system

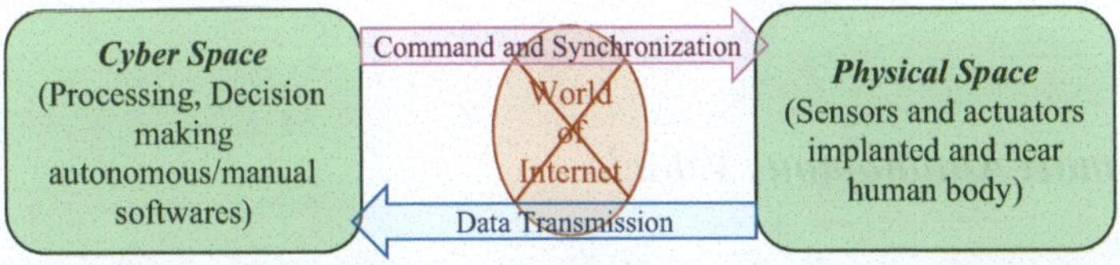

Fig. 8 View of cyber-physical space

(iii) Secure upload, download, and storage of data on a cloud platform using an encryption–decryption approach.
(iv) The most important is to share feedback signals to the inbuilt actuators.

Cyberspace over the world of Internet includes processing, and decision-making capability may send messages to control and coordinate the physical world, i.e., sensors and actuators. The data collection, processing, communication, and control are demonstrated in Fig. 8 [37]:

Medtronic device integration for patient physiological metrics monitoring and sensing is making smart e-health service applications more flexible [38, 39]. Attaching the IoT concept converts the traditional healthcare system into the Internet of Medical Things (IoMT) [40].

Using PUF for device integrity in sophisticated electronic systems is known as hardware-assisted security. Instead of integrating the PUF module with each IoMT, PUFchain 2.0 [41] security primitively offers a scalable alternative by enabling devices to connect and get PUF keys from the edge server. Once the miner has linked the PUF key to a specific media access control (MAC) address, it cannot be linked to any other devices and will remain unique for that MAC address. Variations in the internal micromanufacturing process during chip creation are the basis for

PUF development. Further, the authors presented PUFchain 3.0 [42], a hardware-assisted distributed ledger for robust authentication in healthcare cyber-physical systems.

Due to the dramatic decrease in hospital and clinic visits caused by the COVID-19 epidemic, doctors' offices and clinics are forced to look into online telemedicine and home-monitoring options. A hardware-assisted deep learning technique for low-power detection of cardiovascular abnormalities in smart wearables is presented by Catalani et al. [43].

A machine learning-enabled cyber-physical system is being explored for cyber-attacks in health care [44]. The characteristic of cyber behavior is predicted by the machine learning model and transferred to the medical professionals for the decision-making process. The authors claimed above 96% attack prediction ratio of around 98% accuracy. As per the experimentation, the data transfer delay is reduced by 21%, and communication cost is reduced by 18%. The security of data being transferred over the Internet world is measurably enhanced.

There are a lot many other solutions presented by the research community, but still, a convincing methodology needs to be searched. The devices implanted internally or wearables if compromised can cost a lot to the users.

3.2 Smart Autonomous Vehicle

Self-driving cars or trucks that never require human drivers to take control and ensure the vehicle and user safety are considered smart autonomous vehicles. Figure 9 presents the pictorial demonstration of the autonomous vehicle system.

Fig. 9 Autonomous vehicle demonstration using IoT

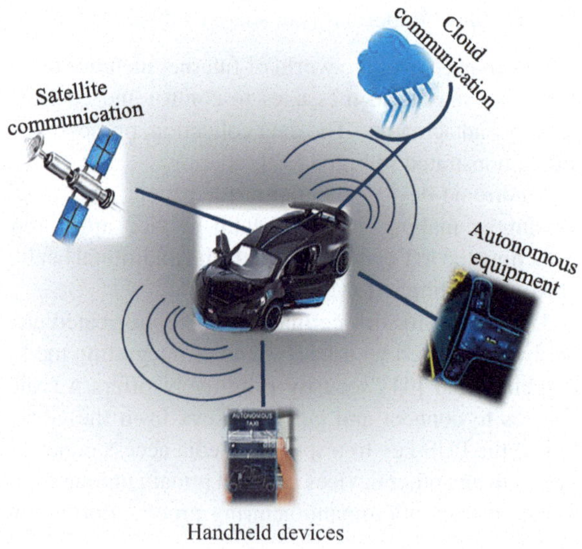

The recent development of IoT devices and smart electronic chips power vehicles not only to handle control systems but also to process and analyze web content on wheels. Due to this reason, vehicles are very much prone to be compromised by unauthorized persons. The security preventions taken during the system design phase are more beneficial than the afterward measures. A lot of work needs to be carried out in this direction. However, some of the existing literature presents the need and initial solution approaches in the domain of autonomous vehicles. Fig. 10 shows how the real-world data is fed to the control unit and further commands the actuators for the action.

Detailed security issues and solutions in automated driving assistants are presented in Ref. 45. Vulnerabilities in the physical components, such as sensors, CPUs, or communication modules, are the main target of hardware assaults in autonomous vehicles. Henceforth, adversarial attacks temper the machine learning algorithms used in AVs. By taking advantage of flaws, AI makes decisions meanwhile. These attacks seek to mislead or impair the operation of AI-based systems, leading to misclassification, jeopardizing system integrity, and endangering user safety.

A hardware-assisted security in electronic control units (ECUs) is proposed by Kornaros et al. [46]. The authors experimented with secure automotive communications by a one-time-programmable network on chip and firewalls.

Configurable hardware-assisted security extension (CHASE) [47] is a unified runtime-configurable hardware-assisted security extension that supports several runtime security settings. CHASE can be customized at runtime based on the desired security level, the real-time availability, and the functional need of the system. This allows for the calibration of the security vs. performance trade-off.

In-vehicle communication and autonomous vehicle security issues are discussed in [48]. Vehicle-to-everything (V2X) communication includes vehicle to vehicle (V2V), vehicle to pedestrian (V2P), vehicle to device (V2D), and vehicle to grid (V2G), and what not in the list to facilitate information sharing to the road infrastructure users in real time. Yadav et al. [49] explored sensor circuits for NBTI (negative bias temperature instability) and PBTI (positive bias temperature instability) resilient SRAM (static random-access memory) design for an autonomous car.

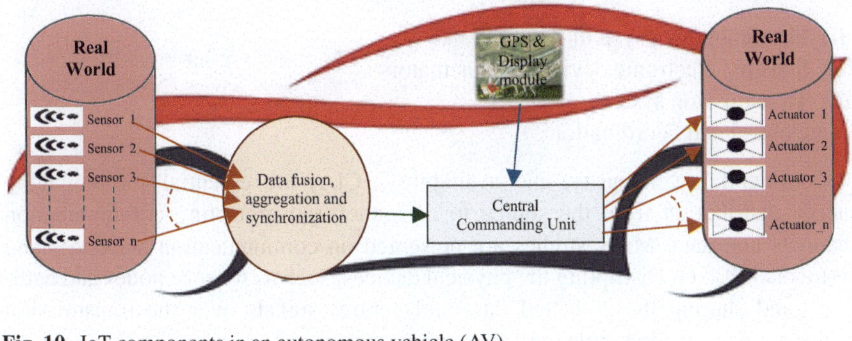

Fig. 10 IoT components in an autonomous vehicle (AV)

Process variation (PV), bias temperature instability (BTI), and time-dependent dielectric breakdown (TDDB) all play crucial roles in semiconductor reliability.

Autonomous vehicles are one of the prominent areas of search for the research community and industry too. A feasible solution for such an automated vehicle needs to be designed so that the user can trust the security mechanism. Nowadays, many big companies, such as Google, Microsoft, Tesla, etc., are working in this direction for a long time and trying to find solutions that ensure users' safety as a whole.

3.3 Smart Grid

A smart grid is an electrical grid equipped with automation, communication, and information technology systems that can track the flow of power from sites of generation to points of consumption and adjust the flow of electricity or reduce the load in real time or almost real time to match the generation.

The information flow over the fast, dependable, and secure data communication network is a crucial component of the smart grid since it allows for the efficient and intelligent management of intricate power systems. The smart grid is undoubtedly becoming more and more significant in the electric power system industry, especially given the current energy sector trends like renewable energy and vehicle electrification. A vital component of smart grid security is secure device-to-device communication. A home network will be used in the future smart home environment to facilitate communication among various heterogeneous smart appliances, devices, and energy management systems. The purpose of smart grid solutions is not only to provide convenience and flexibility to the consumers but also to ensure the security breach in any case.

Fig. 11 shows various components of the smart grid system.

Channelization of power between the generator and consumer is clutched by a smart grid system (SGS) using power electronic converters. Further, the cyber-physical security system (CPS) is coupled with communication and information technology to facilitate operations without compromising transferred data. CPS of the smart grid system is divided into four groups as follows:

(i) Physical power system
(ii) Sensors, electronic devices, and actuators
(iii) Transmission system
(iv) Control and coordination

Confidentiality, integrity, and availability (CIA) are essential for managing energy distribution from the source to consumer and protecting communication channels and data. Many studies are presented on communication standards and protocols [50, 51]. Disrupting the physical devices, such as sensory nodes and actuators, and sharing the collected data and control signals over the transmission medium play a vital role in any smart system. And here is the role of

Fig. 11 IoT-based infrastructure for smart grid

hardware-assisted security mechanisms along with the software-based security approaches. Before granting access for collected data and commands, the system must be smart enough to authenticate the requester.

The block diagram of various components of CPS in SGS is shown in Fig. 12. It shows how sensory nodes and actuators are connected with the electric generator and consumers are connected in the physical space, whereas the generated data are communicated to the control module through the internetworking for further processing, analysis, and so forth command.

However, DoS, FDI, and Man-In-Middle attacks are very common and need to be seriously attended to avoid any blackout. Increasing the resilience of power grid system is done by integrating the cyber-physical security system and cyber security as a single unit [51].

To achieve mutual authentication, each device has to generate and share a private key mechanism that can be compromised. To address such issues, a hardware-assisted solution is proposed in [52]. The authors have created and demonstrated a system that uses Trusted Computing technologies to ensure hardware security for private keys on devices. It is claimed that the private key is only accessible on a particular device and in a secure, trusted environment by utilizing DRTM late-launch functionality.

A secure device authentication scheme is discussed by Sideris et al. [53] for securing message sharing between the communicating devices. The authors discussed the various challenges of security in the smart grid and provided a wide literature addressing data anonymization.

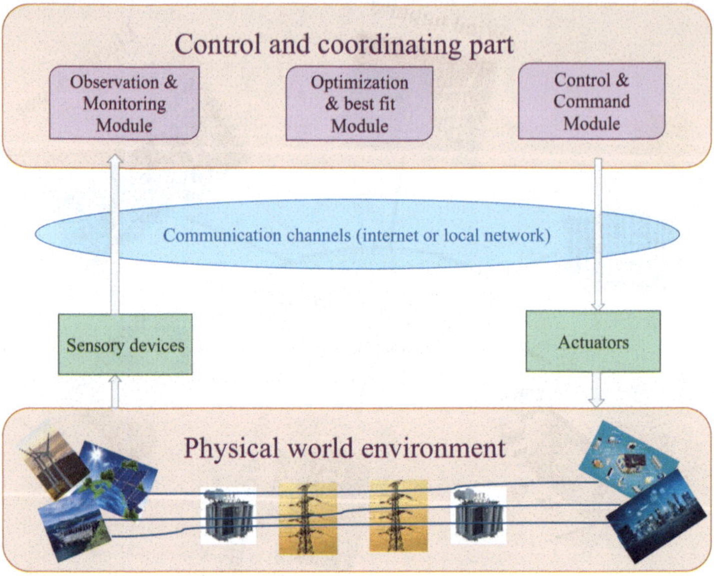

Fig. 12 Block view of CPS in SG

In smart grids, the advanced encryption standard (AES) algorithm enabled by field-programmable gate array (FPGA) technology is used to address various security issues [54]. A testbed is deployed using AES encryption and decryption chip and Xilinx ISE 14.2 software, synthesized on SPARTAN-3E FPGA, and validated through a few test cases to confirm AES's suitability of secure message exchange in smart grid security.

4 Trade-off Between Security and Various IoT Parameters

Difficulties with security principles, cost, implementation issues, and end-user ease must all be balanced with security. Every solution has some trade-offs between the level of security and comfort to the users. Finding the ideal balance between the necessary security levels, implementation costs, and even viability is a measure concern of the smart world. For instance

(i) Some IoT devices require specific sensors and might not have enough processing power to perform sophisticated cryptographic operations. However, all the things described above depend on the application of the IoT system and must be taken care of. In most cases, especially in IoT, the resources are plenty enough to work out the challenges and limitations of resource crisis.
(ii) End-to-end message latency and messages if compromised cannot be affordable, which may lead to disastrous outcomes.

(iii) Energy requirements during the various secure operations may require an alternate second power source for uninterrupted services.

Depending on the applications, use case scenarios, and environment parameters, the required level of confidentiality, integrity, availability, and authentication will always be assessed over various frameworks.

5 Conclusion and Future Directions

As the demand for smartness and digital solutions for real-life objects is increasing, hardware-assisted solutions are essential in the current era. Extensive research work in various domains of the smart world is yet to be carried out for concrete and reliable solutions. Some solutions that have been proposed so far by the research community are presented in this chapter.

The highlights of the chapter can be summarized as follows:

(i) Among embedded security methods, embedded hardware security can safeguard each device's identity and stop unauthorized changes. Such devices are made to operate and safeguard the confidentiality and privacy of the enormous amounts of data these devices produce.
(ii) So far, smart chip security is based on addressing how the device behaves in normal circumstances; however, it must also deal with how to respond to cyberattacks and other tampering, including self-destruction.
(iii) For the billions of linked IoT devices, appropriate security mechanisms can be provided by implementing embedded security techniques, including hardware-based ones, as has already been demonstrated, and should be forwarded into practice for other security use cases.

The research community is working on the various prototypes of solution approaches. However, a lot needs to be taken care of in the future so that guaranteed solutions in the context of security and safety can facilitate the people of our society. Some of the research domains are mentioned below:

(i) Hardware-assisted security may increase the level of security that ensures the safety and data breach of the users. So, lightweight hardware devices in terms of computational power, size, battery requirement, etc. need to be researched or engineered.
(ii) The message exchange should be taken care of by the software security mechanisms because the message on the air is more prone to be compromised.
(iii) Most important is if the user device or message is compromised, it must be dealt with the appropriate action at the earliest so that the loss can be minimized. Such kind of mechanisms and devices need to be developed.
(iv) Hardware-assisted virtualization technology has recently garnered much attention in finding such security solutions. Some big industries, such as Intel,

ARM, AMD, etc., are working in the same direction, and a lot of optimization needs to be explored.

(v) Another approach is trusted computing (TC) for ensuring security over data-in-use, and the area must be explored in the direction of hardware-enabled security systems.

The research community and industry people are continuously working in the direction of a safe and careful secure system for society. Some of the banks, such as HSBC, provide a physical device to each of its users to generate a random number (similar to OTP) to log the bank account/credit card information through the Internet banking system. Such kinds of devices are required for IoT systems.

References

1. Alliance, A. S. C. (2016). Embedded hardware security for IoT applications. In *A smart card alliance internet of things security council white paper.*
2. Hanninen, I., Lu, H., Blair, E., Lent, C., & Snider, G. (2014). *Field-coupled nanocomputing: Paradigms, progress, and perspectives.* Springer.
3. Timler, J., & Lent, C. S. (2002). Power gain and dissipation in quantum-dot cellular automata. *Journal of Applied Physics, 91*(2), 823–831.
4. Jin, Y. (2019). Towards hardware-assisted security for IoT systems. In *2019 IEEE computer society annual symposium on VLSI (ISVLSI)* (pp. 632–637). IEEE.
5. Walter, M., Wille, R., Große, D., Torres, F. S., & Drechsler, R. (2018). An exact method for design exploration of quantum-dot cellular automata. In *2018 design, automation & test in europe conference & exhibition* (pp. 503–508). IEEE.
6. Suh, G. E., & Devadas, S. (2007). Physical unclonable functions for device authentication and secret key generation. In *Proceedings of the 44th annual design automation conference* (pp. 9–14). DAC 2007.
7. Lee, J. W., Lim, D., Gassend, B., Suh, G. E., Van Dijk, M., & Devadas, S. (2004). A technique to build a secret key in integrated circuits for identification and authentication applications. In *2004 symposium on VLSI circuits. Digest of technical papers (IEEE Cat. No. 04CH37525)* (pp. 176–179). IEEE.
8. Azhar, M. J., Amsaad, F., & Köse, S. (2018). Duty-cycle-based controlled physical unclonable function. *IEEE Transactions on Very Large Scale Integration (VLSI) Systems, 26*(9), 1647–1658.
9. Gope, P., & Sikdar, B. (2018). Lightweight and privacy-preserving two-factor authentication scheme for IoT devices. *IEEE Internet of Things Journal, 6*(1), 580–589.
10. Yanambaka, V. P., Mohanty, S. P., Kougianos, E., & Puthal, D. (2019). PMsec: Physical unclonable function-based robust and lightweight authentication in the internet of medical things. *IEEE Transactions on Consumer Electronics, 65*(3), 388–397.
11. Zhao, B., Zhao, P., & Fan, P. (2020). ePUF: A lightweight double identity verification in IoT. *Tsinghua Science and Technology, 25*(5), 625–635.
12. Maiti, A., Casarona, J., McHale, L., & Schaumont, P. (2010). A large scale characterization of RO-PUF. In *2010 IEEE international symposium on hardware-oriented security and trust (HOST)* (pp. 94–99). IEEE.
13. Shi, J., Lu, Y., & Zhang, J. (2019). Approximation attacks on strong PUFs. *IEEE Transactions on Computer-Aided Design of Integrated Circuits and Systems, 39*(10), 2138–2151.

14. Amsaad, F., Oun, A., Niamat, M. Y., Razaque, A., Kose, S., Mahmoud, M., Alasmary, W., & Alsolami, F. (2021). Enhancing the performance of lightweight configurable puf for robust iot hardware-assisted security. *IEEE Access, 9*, 136792–136810.
15. Amsaad, F., Razaque, A., Baza, M., Kose, S., Bhatia, S., & Srivastava, G. (2021). An efficient and reliable lightweight PUF for IoT-based applications. In *2021 IEEE international conference on communications workshops (ICC workshops)* (pp. 1–6). IEEE.
16. Maiti, A., & Schaumont, P. (2009). Improving the quality of a physical unclonable function using configurable ring oscillators. In *2009 international conference on field programmable logic and applications* (pp. 703–707). IEEE.
17. Dofe, J., Frey, J., & Yu, Q. (2016). Hardware security assurance in emerging IoT applications. In *2016 IEEE international symposium on circuits and systems (ISCAS)* (pp. 2050–2053). IEEE.
18. Kornaros, G. (2022). Hardware-assisted machine learning in resource-constrained IoT environments for security: Review and future prospective. *IEEE Access, 10*, 58603–58622.
19. Trusted Base System Architecture Client. (2018). (4th ed) System Hardware on ARM Document Number: ARM Den 0021D.
20. ARM Limited. (2019). ARMv8-M architecture reference manual.
21. https://www.intel.com/content/www/us/en/business/enterprise-computers/resources/trusted-platform-module.html#:~:text=A%20TPM%2C%20or%20a%20trusted,motherboard%20or%20in%20its%20processor
22. ISO/IEC, ISO/IEC 11889:2015 Trusted Platform Module. (2015).
23. Fremantle, P., & Scott, P. (2017). A survey of secure middleware for the internet of things. *PeerJ Computer Science, 3*, e114.
24. Humayed, A., Lin, J., Li, F., & Luo, B. (2017). Cyber-physical systems security—A survey. *IEEE Internet of Things Journal, 4*(6), 1802–1831.
25. Kornaros, G., Wozniak, E., Horst, O., Koch, N., Prehofer, C., Rigo, A., & Coppola, M. (2018). Secure and trusted open cps platforms. In *Solutions for cyber-physical systems ubiquity* (pp. 301–324). IGI global.
26. Nguyen, T. T., & Reddi, V. J. (2021). Deep reinforcement learning for cyber security. *IEEE Transactions on Neural Networks and Learning Systems, 34*(8), 3779–3795.
27. de Araujo-Filho, P. F., Kaddoum, G., Campelo, D. R., Santos, A. G., Macêdo, D., & Zanchettin, C. (2020). Intrusion detection for cyber–physical systems using generative adversarial networks in fog environment. *IEEE Internet of Things Journal, 8*(8), 6247–6256.
28. Deokar, B., & Hazarnis, A. (2012). Intrusion detection system using log files and reinforcement learning. *International Journal of Computer Applications, 45*(19), 28–35.
29. Patel, N., Sasan, A., & Homayoun, H. (2017). Analyzing hardware based malware detectors. In *Proceedings of the 54th Annual Design Automation Conference* (pp. 1–6).
30. Tang, A., Sethumadhavan, S., &Stolfo, S. J. (2014, September 17–19). Unsupervised anomaly-based malware detection using hardware features. In *Research in attacks, intrusions and defenses: 17th international symposium, RAID 2014*. pp. 109–129, Springer.
31. Wang, X., & Karri, R. (2015). Reusing hardware performance counters to detect and identify kernel control-flow modifying rootkits. *IEEE Transactions on Computer-Aided Design of Integrated Circuits and Systems, 35*(3), 485–498.
32. Theissler, A. (2017). Detecting known and unknown faults in automotive systems using ensemble-based anomaly detection. *Knowledge-Based Systems, 123*, 163–173.
33. Pachauri, G., & Sharma, S. (2015). Anomaly detection in medical wireless sensor networks using machine learning algorithms. *Procedia Computer Science, 70*, 325–333.
34. Liu, Y., Garg, S., Nie, J., Zhang, Y., Xiong, Z., Kang, J., & Hossain, M. S. (2020). Deep anomaly detection for time-series data in industrial IoT: A communication-efficient on-device federated learning approach. *IEEE Internet of Things Journal, 8*(8), 6348–6358.
35. Sengupta, A., & Kundu, S. (2017). Guest editorial securing IoT hardware: Threat models and reliable, low-power design solutions. *IEEE Transactions on Very Large Scale Integration (VLSI) Systems, 25*(12), 3265–3267.

36. Dong, C., He, G., Liu, X., Yang, Y., & Guo, W. (2019). A multi-layer hardware trojan protection framework for IoT chips. *IEEE Access, 7*, 23628–23639.
37. Verma, R. (2022). Smart city healthcare cyber physical system: Characteristics, technologies and challenges. *Wireless Personal Communications, 122*(2), 1413–1433.
38. Sundaravadivel, P., Kougianos, E., Mohanty, S. P., & Ganapathiraju, M. K. (2017). Everything you wanted to know about smart health care: Evaluating the different technologies and components of the internet of things for better health. *IEEE Consumer Electronics Magazine, 7*(1), 18–28.
39. Sun, J., Khan, F., Li, J., Alshehri, M. D., Alturki, R., & Wedyan, M. (2021). Mutual authentication scheme for the device-to-server communication in the internet of medical things. *IEEE Internet of Things Journal, 8*(21), 15663–15671.
40. Razdan, S., & Sharma, S. (2022). Internet of medical things (IoMT): Overview, emerging technologies, and case studies. *IETE Technical Review, 39*(4), 775–788.
41. Bathalapalli, V. K., Mohanty, S. P., Kougianos, E., Baniya, B. K., & Rout, B. (2022). Pufchain 2.0: Hardware-assisted robust blockchain for sustainable simultaneous device and data security in smart healthcare. SN computer. *Science, 3*(5), 344.
42. Bathalapalli, V. K., Mohanty, S. P., Kougianos, E., Iyer, V., & Rout, B. (2024). PUFchain 3.0: Hardware-assisted distributed ledger for robust authentication in healthcare cyber–physical systems. *Sensors, 24*(3), 938.
43. Catalani, A., Chatzigiannakis, I., Anagnostopoulos, A., Akrivopoulou, G., Amaxilatis, D., & Antoniou, A. (2021). Hardware-assisted and deep-learning techniques for low-power detection of cardiovascular abnormalities in smart wearables. In *2021 IEEE international conference on smart internet of things (SmartIoT)* (pp. 144–151). IEEE.
44. AlZubi, A. A., Al-Maitah, M., & Alarifi, A. (2022). Cyber-attack detection in healthcare using cyber-physical system and machine learning techniques. *Soft Computing, 25*(18), 12319–12332.
45. Mehta, A., Padaria, A. A., Bavisi, D., Ukani, V., Thakkar, P., Geddam, R., Kotecha, K., & Abraham, A. (2023). Securing the future: A comprehensive review of security challenges and solutions in advanced driver assistance systems. *IEEE Access, 12*, 643.
46. Kornaros, G., Tomoutzoglou, O., & Coppola, M. (2018). Hardware-assisted security in electronic control units: Secure automotive communications by utilizing one-time-programmable network on chip and firewalls. *IEEE Micro, 38*(5), 63–74.
47. Dessouky, G., Zeitouni, S., Ibrahim, A., Davi, L., & Sadeghi, A. R. (2019). CHASE: A configurable hardware-assisted security extension for real-time systems. In *2019 IEEE/ACM international conference on computer-aided design (ICCAD)* (pp. 1–8). IEEE.
48. Shon, T. (2021). In-vehicle networking/autonomous vehicle security for internet of things/ vehicles. *Electronics, 10*(6), 637.
49. Yadav, N., Kim, Y., Alashi, M., & Choi, K. (2020). A reactive and on-Chip sensor circuit for NBTI and PBTI resilient SRAM design. *Electronics, 9*, 326.
50. Jha, A. V., Appasani, B., Ghazali, A. N., Pattanayak, P., Gurjar, D. S., Kabalci, E., & Mohanta, D. K. (2021). Smart grid cyber-physical systems: Communication technologies, standards and challenges. *Wireless Networks, 27*(4), 2595–2613.
51. Abrahamsen, F. E., Ai, Y., & Cheffena, M. (2021). Communication technologies for smart grid: A comprehensive survey. *Sensors, 21*(23), 8087.
52. Paverd, A. J., & Martin, A. P. (2012). *Hardware security for device authentication in the smart grid. In international workshop on smart grid security* (pp. 72–84). Berlin, Heidelberg, Springer.
53. Sideris, A., Tsiktsiris, D., Ziouzios, D., & Dasygenis, M. (2019). Smart grid hardware security. In *IoT for Smart Grids: Design Challenges and Paradigms* (pp. 85–113). Springer.
54. Kumar, N., Mishra, V. M., & Kumar, A. (2020). Smart grid security with AES hardware chip. *International Journal of Information Technology, 12*, 49–55.

Secure Supply Chain and Manufacturing

K. Sudeendra Kumar, S. S. Rekha, Ayas Kanta Swain, and K. K. Mahapatra

1 Introduction

It is difficult to define Internet of Things (IoT) precisely. Internet of Things (IoT) is a network of heterogeneous objects, where the object may be a sensor or any electronic gadget. The idea of IoT emerged in the last decade due to advances in semiconductor and computer technologies, which have a capacity to bring significant change in almost every economic activity: agriculture, manufacturing, and services. The IoT refers to the networking of objects through the Internet to deliver the products and services. The IoT ecosystem includes embedded systems, data analytics, cloud computing, networking, sensors, and actuators. Availability of high-performance computing elements (processors) at low cost, advances in storage techniques (cloud), and affordable wireless communication systems (4G, 5G) to connect the objects are the factors responsible for making the idea of IoT into reality. Several IoT reference models have been discussed in academic and industrial literature. A comprehensive seven-layer model described by Cisco Systems is a widely accepted IoT model [1]. Cisco Systems' suggested model is shown in Fig. 1. The seven layers of the Cisco model are: Level 1 (Edge devices): The edge nodes are also known as perception layer devices which collect data from sensors and interact with physical objects in the ecosystem. The edge nodes are generally powered by microcontrollers/SoC and connected to Internet through gateways. Level 2 (Communication): This layer consists of components that enable transmission of information between the devices at the edge layer and information exchange between the edge layer and

K. Sudeendra Kumar (✉) · S. S. Rekha
PES University, Bangalore, India
e-mail: sudeendrakumark@pes.edu; rekha.ss@pes.edu

A. K. Swain · K. K. Mahapatra
National Institute of Technology, Rourkela, India
e-mail: swain.ayas@nitrkl.ac.in; kkm@nitrkl.ac.in

© The Author(s), under exclusive license to Springer Nature Switzerland AG 2025

A. Mishra et al. (eds.), *Hardware Security: Challenges and Solutions*,
https://doi.org/10.1007/978-3-031-81213-2_4

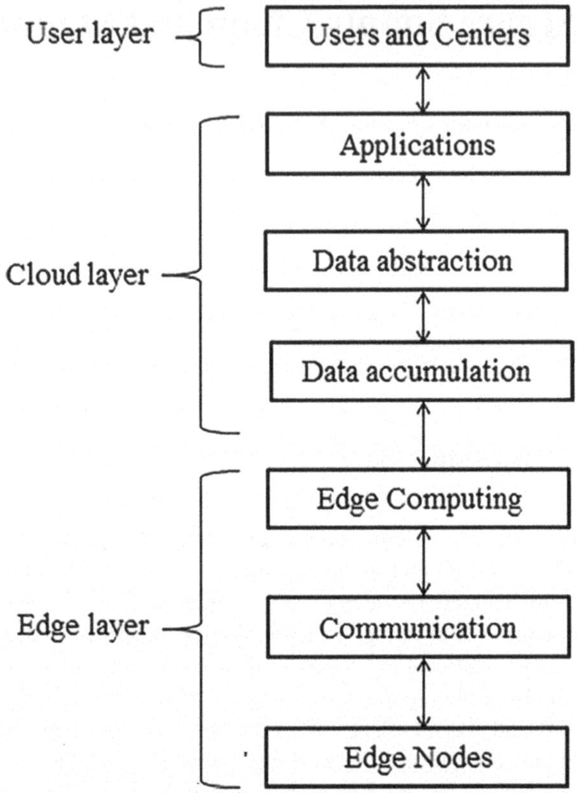

Fig. 1 Cisco internet-of-things seven-layer model

higher layers. Gateways are generally standalone devices, which support wireless and wireline communication protocols. Level 3 (Edge computing): also known as fog computing where the data processing is initiated. The amount of processing depends upon the applications and it is generally limited to simple data crunching. Level 4 (Data accumulation):—This layer is meant to convert the data from one format to the other, e.g., network packets to database tables, etc. Depending on the applications, this layer can also perform data filtering and selective storing. Level 5 (Data abstraction): The layer performs moderate amount of data crunching to make the processing efficient at higher levels. Level 6 (Applications): This layer performs data interpretation for different applications and industrial needs. Level 7 (Users and Centres): Users make use of products and services.

In the IoT ecosystem, most of the gadgets and things are connected to the Internet. Due to this, there is an increase in the number of threats and attacks on security or privacy. The overall security of the IoT ecosystem is complex. The intention of adversary might be stealing sensitive information like passwords, financial accounts, and location data. Other motivations of an adversary include different types of denial of service (DoS) attacks in any layer of IoT hierarchy.

In IoT ecosystem, security design should be more holistic and comprehensive in comparison with general desktop computers or conventional embedded systems. Security of IoT cannot be addressed by installing the software patch or by adding an extra hardware module. The security must be designed layer by layer across the IoT hierarchy. The edge layers (edge nodes) are most vulnerable to attacks, and adversaries can get access to complete network and create catastrophes, as the edge nodes are generally placed in remote locations to collect data and to control physical objects. Securing the edge device is a challenging problem. Generally, edge devices manage sensors and actuators which directly control the physical objects in well-established IoT applications like smart city and smart grid and smart transport. Microcontroller/SoC is an inevitable part of edge node design and most active part of the edge device. SoC plays a significant role in securing the IoT edge node, and SoC device should have security-related modules to support the security of the edge node. The necessary security properties the IoT edge device should have are: hardware-based root of trust, trusted computing base, secure firmware/software update, anti-counterfeiting solution, countermeasures against different types of Side-Channel Attacks (SCAs), and malicious attacks [2].

Hardware and software must work together to implement the security requirements with the IoT edge node security rooted in hardware and guarded by evolving software/firmware. Most of the modern-day microcontrollers/SoC lack features to support these security properties. Few microcontrollers support some of the above-mentioned security properties with cryptographic modules, which are not enough to address all varieties of threats [3].

Due to globalization of semiconductor design and manufacturing, SoC ODMs (Original Design Manufacturers) are facing several hardware security (SoC security) issues like counterfeiting, IP violations, side-channel attacks, and hardware Trojans. The SoC security issues and IoT edge node security issues are tightly connected topics. The solutions developed by SoC manufacturers against hardware security issues can be used to address IoT security issues directly or can be enhanced further to make them useful for IoT edge node security. IoT edge node security and SoC security are interwoven in such a way that the IoT edge node security can be addressed from the beginning of the SoC design process. It is difficult to construct effective security for IoT edge node, if the SoC used to design edge node does not have required security features. A careful thought to hardware security during SoC design will lead to inclusion of security measures to be built into the SoC device which will make the basement for IoT security strong and effective. The intelligent IoT edge device brings tremendous values in the form of IP (Intellectual Property) and monetization. So, intelligent IoT edge device is a target for hacking and demands stringent security. To make IoT edge node more secure, the necessary considerations the SoC/microcontroller makers should take are discussed in a study [2]. A secure microcontroller should address all SoC security issues and support the IoT edge device security. This survey summarizes

- How SoC security and IoT edge node security are tightly connected topics.

- The security features the SoC vendors should include in their chips to make their devices attractive to IoT edge node designers.
- Providing the readers the comprehensive review of both hardware security and IoT edge node security.
- The existing solutions which are designed to address both hardware and edge node security together.
- Necessity of centralized security IP as a holistic solution for both SoC and edge node security and a summary of existing centralized IPs from both industry and academia.
- Finally, a conceptual foundation for Secure Infrastructure IP (SIIP) with features required to address both SoC security and edge node security.

The remainder of the chapter is organized as follows. In Sect. 2, we present the background of hardware security and IoT edge node security. In Sect. 3, we summarize the existing solutions for hardware security and IoT edge node security. Then, in Sect. 4, we describe the necessity of centralized security IP and present the survey of existing centralized IPs. In Sect. 5, we discuss a conceptual foundation for ideal Secure Infrastructure IP (SIIP) with necessary security features, and finally, we conclude in Sect. 6.

2 Hardware Security and IoT Edge Node Security

In this section, we briefly discuss both IoT edge device security and hardware security issues.

2.1 IoT Edge Node Security

The well-known edge computing nodes are wireless sensor nodes and other control nodes. The attacks on edge node are as follows:

Malicious Activity: Malicious modifications in the firmware/software to attack edge nodes are well known [4]. Malicious code injections can leak data or lead to DoS. The taxonomy and more details on firmware Trojans can be found in Ref. [5]. Hardware Trojan is a malicious modification in chip design by an adversary to leak the confidential information or to perform any other kind of security breach like DoS attack, etc. The HT can be inserted into the design during RTL coding, test insertion, or fabrication. An adversary in the design house can insert HT or third party IP core sourced from untrusted vendor that may have hardware Trojan. The detailed taxonomy and survey on HT can be found in Ref. [6]. The presence of hardware Trojan in the microcontroller/SoC used in the edge node will be catastrophic for both the user and the product/service provider.

Different Types of Side-Channel Attacks (SCAs): Side-channel attacks are performed to leak the secret key or confidential information, generally performed on cryptographic cores. SCA is also performed on other sensitive assets of the microcontroller/SoC device to leak IP access keys and trade-related secrets. The power-based SCA is a well-researched technique in which the adversary collects and analyzes the power traces to derive the key used for encryption [7]. Similarly, the timing analysis-based SCA is discussed in [8]. The design for testability (DFT) structures which increase the observability and controllability of the design which is beneficial during testing is misused by an adversary to perform test-based SCA [9]. The electromagnetic (EM) waves emitted by the device can be processed to get valuable information about the working of the device, as presented in [10]. Fault attacks create unexpected environmental conditions and variations in the normal functioning of global signals like clock, etc., to induce faults which reveal the internal states in cryptographic implementations, as presented in [11]. An adversary can also develop multilevel attacks which are a mix of the abovementioned attacks which are more dangerous to both the user and the product/service provider [12]. The widely accepted assumption is that the probability of a successful SCA depends upon the amount of time an adversary will have access to a device. In the case of IoT edge nodes, as they are placed in remote and discrete locations, in most of the cases, an adversary will get unrestricted access to edge nodes which increases the probability of successful SCA.

***Denial of Service (DoS) Type Attacks*:** DoS attacks make the edge device unavailable for its intended users for a temporary period or indefinitely. Different types of DoS attacks on edge nodes are draining out battery, sleep deprivation, and outage attacks. The IoT edge nodes carry small-size batteries with a limited energy capacity. It is easy for an adversary to conduct battery draining attacks by transmitting a large number of random packets to a node as input so that it will run endlessly. This is an example of direct attack through common input channels to perform attacks. Another indirect form of DoS attack described in [13] is called sleep deprivation attack in which battery is the target. In this type of attack, the attacker sends a set of legitimate requests to make the node to run its software loops continuously, which is unfound during the normal operation of the node which leads to battery draining. Other than battery draining attacks, outage attack is another type of attack in which the edge node device stops performing its crucial operations like the ability to detect the catastrophic events which are necessary during emergency [14]. Illegal possession of IoT edge nodes, security compromise in the network gateway, and deploying unauthorized firmware/software on edge nodes lead to most of the DoS-type attacks.

Counterfeit Edge Nodes: Counterfeiting of edge nodes can be categorized as: node replication attacks, malicious nodes, and camouflage nodes. In node replication attack [15], an adversary will add a new node to an existing set of nodes in the network by replicating the identification number of the other genuine node. The adversary may perform a DoS-type attack or leak secret information with the help of a counterfeit node. In camouflaging [16], the adversary inserts the malicious node which acts like a genuine/authorized node and hides the presence of the genuine node. The malicious node injected into the network to obtain access to other nodes is generally intended to leak confidential information.

Apart from the above-discussed security issues, the IoT ecosystem lacks standards in product/service life cycle management. Applications of IoT are so diverse that the development of standards is not easy, and such efforts are in a very nascent stage. Lack of standards in hardware and software development, testing, and deployment procedures can give rise to unknown serious security concerns.

2.2 Hardware Security

In the last decade, hardware security has emerged as one of the important research areas in the field of chip design. Globalization of semiconductor industry has brought changes in design, manufacturing, test, and EDA. The supply chain/service chain is fragmented across different continents, and this has resulted in several hardware security concerns like counterfeiting, hardware Trojans, IP violations, side-channel attacks, and reverse engineering [17]. Hardware security issues are very important, and security is a new tenet in chip design along with performance, power, area, and yield. SoC design companies must perform their activities from design to marketing time-effectively and cost-effectively to remain competitive to the customers. The time to market pressures has pushed SoC vendors to change their operating models. The major changes observed are as follows:

- Depending on the third-party IP (Intellectual Property) vendors for different functionalities, who are specialized in that application domain (e.g., image processors, hardware accelerators for computationally intensive applications, etc.). Both fabless and fab-owned SoC companies depend upon third Party IP (3PIP) cores [18].
- Depending upon contract manufacturers (like TSMC, Global foundries etc.) to manufacture the chips and Outsourced Assembly and Test (OSAT) centers for assembly, packaging, and test. Fabless SoC design houses completely depend upon contract manufacturers and OSAT centers for production operations. Even few fab-owned companies follow the fab-lite models. The companies following fab-lite model will outsource fabrication to contract manufacturers and perform packaging and testing in-house. And there are different forms of fab-lite model companies [19].
- Depending upon the service providers, who involve in design process like verification, post-silicon validation, etc. Service providers also provide specialized manpower or consultants for specific steps in the design process. Both fab-owned and fabless semiconductor companies depend upon these service providers to remain competitive in the market in terms of both time to market and profit margin pressures [18].

The abovementioned changes have led to the fragmentation of SoC design process and manufacturing across geographies. This fragmented business model in both design process and semiconductor supply chain raises the abovementioned hardware security issues [17]. There is a need to have a security framework during chip design which addresses all security issues to prevent revenue and reputation loss to chip makers. We briefly discuss the major hardware security problems:

Hardware Trojans (HT): HT is discussed briefly above. The research connected with HT can be classified as: HT design, HT detection, vulnerability analysis of designs for HT insertion, and defense against HT. The Original Device Manufacturer (ODM) should ensure that there are no HTs in their chips by adopting the suitable security framework at both pre-silicon and post-silicon validation stages [20].

Counterfeit ICs: Counterfeiting and IC piracy are serious threats to the revenue of original component manufacturers. There is a great concern in both industry and government because of its negative impact on innovation, employment, and consumer safety. It is estimated that the counterfeit part's business turnover was around $1.7 trillion in 2015 [21]. The major sources of counterfeiting are: (i) extraction of counterfeit parts from the obsolete electronic equipment/gadgets, (ii) out-of-license overbuilding of chips at untrusted foundries, and (iii) defective or out-of-spec chips coming out from the untrusted OSAT (Out-Sourced Assembly and Test) centers. The ODM must take suitable countermeasures to mitigate the counterfeit chips coming out from the above sources to protect their business interests [19].

Test/Debug Security: Different kinds of side-channel attacks on cryptographic chips are well known [9]. In addition to cryptographic cores, modern processors include trade-related secrets and assets related to Digital Rights Management (DRM) which are crucial for ODM. Debug instrumentation has become mandatory in modern SoC to collect data, which is crucial for diagnosis and failure analysis. The adversary will use test structures and debug instrumentation to leak sensitive information like cryptographic keys and DRM secrets. The ODM must incorporate suitable countermeasures to prevent the attacks based on test and debug modules [9, 22].

***IP Violations/False Claim of Ownership*:** An IP right refers to the intellectual property right of the owner who creates it must be protected to promote innovation. Patents, copyrights, and trademarks are not sufficient to ensure rightful ownership of reusable soft IP cores which are delivered as RTL codes. Security for IP cores can be of two types: (i) protection to IP vendor and (ii) protection to user.

Untrusted SoC integrator procures the IP core from the vendor and may claim its ownership. In such cases, a vendor watermark on the IP core protects the IP vendor during litigation. In the second case, when the vendor sells extra copies of soft IP and blame the SoC integrator, the SoC integrator can include the user watermark above the vendor's watermark to protect his rights [23, 24].

Comparative Analysis: We have briefly discussed IoT edge node security and hardware security. We present the comparative analysis of hardware security and IoT

Table 1 Comparative analysis of hardware security and IoT edge device security

Hardware security	IoT edge device security	Solutions designed to address hardware security issues can be scaled to address IoT edge device security?
Hardware Trojan (HT)	Malicious activity	IoT designers have no role in the HT-related threats, and it is ODM to ensure that chips are HT-free. Microcontroller/SoC should support run-time defense mechanisms to protect sensitive assets against malicious activity. Battery draining-based DoS attacks can be due to HT or software/firmware running on the IoT edge node. The measures incorporated by ODM to defend HT attacks can be used to address malicious activity-based attacks on IoT nodes.
SCA and test/debug security	Different types of SCA.	The measures designed for debug security and SCA attacks can be directly used or scaled to address the challenges in IoT edge device security.
Counterfeit parts	Counterfeit nodes	The anticounterfeiting techniques like hardware metering designed by SoC vendors can be scaled up to address the issue of counterfeit IoT nodes.
Hardware IP violations	IP violations in IoT edge node	Hardware metering-based hardware IP protection schemes can be used to address the IoT edge-level IP security issues.

edge node security in Table 1. From Table 1, it is evident that SoC security and IoT edge node security are deeply connected topics, and it is difficult to address the IoT edge node security issues without the support of the SoC designer. And most of the solutions designed for hardware security problems (which directly affect the revenue and reputation of chip makers) can be used directly or scaled up to address the security problems of IoT edge nodes.

3 Existing Solutions

We will discuss the existing techniques of hardware security which are scaled to IoT edge node security in this section.

3.1 Counterfeiting

Counterfeit ICs are also not reliable and catastrophic when used in critical applications. To address the problem of counterfeit parts, IC industry has come up with a technique called IC/Hardware metering. Hardware metering was first coined in 2001 [25] in which every IC will carry a unique tag to identify itself. The development of Physical Unclonable Functions (PUFs) has led to more efficient active IC metering schemes in the last decade. IC manufacturers, mainly microcontroller/SoC design houses, adopt PUF-.

based metering techniques to address the menace of counterfeit ICs. The PUF-based anticounterfeiting schemes prevent the counterfeit parts coming out from untrusted foundries and OSAT centers [19, 78]. PUFs produce nondeterministic responses/signatures for a given challenge (input) based on inherent random variations occurring during the manufacturing process of chips. The input (challenge) to the PUF and its corresponding output (response) is called CRP (Challenge–Response Pair). PUF can generate response on the fly which overcomes the disadvantages of conventional methods of storing keys in nonvolatile memories. Initial versions of PUF designs had reliability issues like variations in CRP due to the change in temperature and aging effects.

Recent PUF designs are more stable and reliable, which can be used in hardware metering applications by chip makers. PUF is not only useful in hardware metering applications to fight IC counterfeiting but also can be used in IoT edge device identification and authentication. Anticounterfeiting schemes for IoT edge device must be tamper-resistant, lively, and reliable. There are several IoT edge device identification techniques like Samsung KNOX [26], Intel SGX [27] (Software Guard Extension), and ATRIUM [28] that need modifications in processors which are not accepted by most of the SoC manufacturers under the present IP licensing regime. Hardware-based attestation and identification approaches are attractive to SoC manufacturers, and PUF fits the requirement. Several PUF-based anticounterfeiting solutions designed for IoT edge node are briefly discussed: -

ReSC: RFID-enabled supply chain management, traceability, and attestation scheme (ReSC) presented in [29] is a promising solution against counterfeit nodes, split attacks, and product theft. ReSC-enabled IoT edge node consists of a read-only RFID tag stored in locked memories, a control chip that includes SRAM PUF, and a wireless communication module. The RFID tag and PUF-generated chip identity prevent split attacks and the insertion counterfeit node through one-to-one mapping of edge nodes. SRAM PUF is used along with RFID tags as an anticounterfeiting solution for IoT edge node identification and authentication.

PUF for FPGA Bitstream Update: FPGA devices used in IoT applications can make use of dynamic partial reconfiguration (DPR) to update the functionality. This feature can be misused to update the FPGA with malicious bitstream which will compromise the security of the IoT edge node. The authors prove that the PUF-based security mechanism for online bitstream update provides better security than conventional cryptographic techniques. PUF-based security protocols used for identifying the genuine node for online software/firmware update is more secure than conventional protocols. PUF-based protocol defends against both malicious update and the insertion of the counterfeit node [30].

PUF-Based IoT Edge Node Authentication: The PUF-based edge node authentication proposed in [31] presents a mechanism which does not require the storage of challenge–response pairs (CRPs) at the verifier. The key exchange protocol is designed by combining the Identity-Based Encryption (IBE) and PUF. The IoT edge node includes PUF and generates the responses for a given challenge. The response is verified by the verifier node which is placed at higher layers of the IoT ecosystem. The verifier runs a PUF model to generate the response to compare it with the response received from the edge node for authentication.

In most of these protocols, the edge nodes are verified for their integrity and identification by verifier nodes. The research on PUF is not limited to academics, and there are several case studies which show that industry is also embracing the PUF for security applications. The DeepCover security solution in Maxim microcontroller MAXQ1061 uses PUF to generate the cryptographic keys [32]. Intel Stratix 10 FPGA SoC devices consist of Security Device Manager (SDM) [33] which includes PUF as part of the device configuration process for key generation and device identification. SRAM PUF technology developed by intrinsic ID is used in SDM [34]. The PUF module consists of an algorithmic component to address the aging issues. The algorithmic component (software) can be updated so that the PUF circuit can be used for the lifetime of the FPGA device.

Similarly, Xilinx Zynq Ultrascale FPGA devices carry hardened PUF to produce the device-specific unique signature [35]. In earlier FPGA devices, the bitstream is decrypted using on-chip AES decryption engine and AES key saved in on-chip non-volatile memory, which is vulnerable to attacks. In Zynq devices, PUF is used to create the Key Encryption Key (KEK), which is used to encrypt the key, which makes the bitstream update and authentication more secure.

We have discussed recent techniques useful to mitigate the counterfeit ICs and counterfeit nodes in the IoT edge layer. Conclusions from the above discussion are as follows:

- PUF is a promising security primitive useful to address both IC counterfeiting and IoT edge node counterfeiting.
- The active component of the IoT edge node (SoC/microcontroller) must have an in-built PUF, which is useful in both SoC security and IoT edge node security as an anticounterfeiting solution.

3.2 Test/Debug Security

Systems-on-Chips (SoCs) designed in modern times are complex in terms of functionality, circuit density, and frequency of operation. These complexities have led to new challenges in post-silicon validation. The scan chains are common test structures which are inserted into the design to make testing easy and efficient. The recent advancements in post-silicon debugging can find highly sophisticated errors and bugs by incorporating the on-chip instruments. The well-known instruments are embedded logic analyzers, trace buffers, and instruction recorders to store the states in the processor microarchitecture. To guarantee the reliability of the chip and to further analyze the failures in chips, instruments for in-field monitoring, diagnosis, error correction, and repair are included in the chip. Regular assertion checkers to monitor malfunction, silicon odometers to monitor aging, and workload monitors are placed at appropriate locations to collect data. These data traces can be used to analyze the health of a chip on-field and also for failure analysis in case the chip fails in field. Debug instrumentation has become mandatory in modern SoC to collect data, which is crucial for diagnosis at the time of failure, both in production testing and during in-field diagnosis [36].

The design for testability (DFT) or scan chain-based side-channel attacks was first reported in [37]. In the last one decade, there are several types of DFT-based SCA's and countermeasures proposed. Apart from DFT-based SCA, there are several other vulnerabilities connected with the test and debug structures that exist in modern-day SoC devices. The IEEE 1500 core test standard [38] and on-chip instrument network standard IEEE P1687 [39] increase the observability and controllability of the internal states of SoC which is a threat to security. Adversaries can leak cipher keys and other sensitive trade-related information using advanced debug structures. The primary debug interface to most of the on-chip debug instruments is JTAG. The JTAG security mechanisms proposed in the past are based on password-based access restriction and data obfuscation schemes [40]. There are several ad hoc test/debug security mechanisms for test standards like IEEE 1500 and IEEE P1687 (IJTAG) that are discussed in [41, 42]. An ad hoc scheme does not serve the security requirement of SoC test/debug infrastructure, and a centralized architecture which addresses the security of different test standards and networks is required to address the test security problem more holistically. There are other several techniques to prevent the test-based side-channel attacks. The paper [79] propose a dynamic-key-based secure scan architecture that works in tandem with Physical Unclonable Function (PUF) which defend the scan-based attacks while facilitating both manufacturing and in-field testing. The SST techniques are further improved by incorporating the Return Material Authorization (RMA) capability, and lightweight ciphers used in SST make it more efficient in terms of test power consumption and test time [80]. The CHSM (Chiplet Hardware Security Module) proposed in [81] includes several types of sensors to address counterfeiting and also includes the IP cores for cryptographic acceleration. The CHSM described in [81] is a comprehensive solution for high-end System-on-Chip devices which include crypto-accelerators and secure memory, where such chips are targeted for personalized gadgets like mobile phones, etc. There is a requirement for lightweight security IP which addresses counterfeiting, run-time defense against hardware Trojans, and holistic debug security for low-end edge nodes widely used in sensor network applications.

Both test and security are equally important, and a comprehensive solution should be as follows:

- Centralized architecture: Different test and debug infrastructure used at different phases of SoC device life cycle must be protected by comprehensive policy decisions which strike a balance between debug and security requirements. The centralized comprehensive policy decision supports different test standards used in the modern-day SoC design.

- Late Changes: Most of the modern complex SoC designs will undergo two to three re-spins to meet the design specifications. The security policy must be flexible enough to adopt late changes in the test/debug architecture.
- Reusability: The reusable systemic centralized debug architecture is an essential solution.

The centralized debug architecture which is an essential requirement for the SoC device to protect its crucial assets can be used directly to prevent test-based side-channel attacks on IoT edge nodes and can be scaled to address the IoT edge security issues by incorporating the following:

- Secure boot and secure firmware/software upgrade.
- Monetizing hardware and software IP cores.

3.3 Power-Based SCA

SoC designers can prevent power-based SCA by incorporating suitable countermeasures. Modern SoCs incorporate structured low-power schemes like power gating, clock gating, and DVFS (Dynamic Voltage Frequency Scaling) techniques. These low-power schemes can be used as countermeasures against power-based SCA [43]. In most of the advanced SoC devices, power management sections are programmable, and countermeasures against power-based SCA can be made user-programmable to obfuscate the power traces which will confuse the adversary. The programmable power management module can be used to prevent SCA attacks on the IoT edge node also.

3.4 Hardware Trojan/DoS Attacks

A rogue employee in the design house may insert malicious code into the design, or third-party IP core may have malicious code which is known as hardware Trojan, which makes the system fail during critical function in operation or leak confidential information. The ODM should ensure no malicious code or hardware Trojan is inserted in the chip. Firmware supplied from a third-party hardware IP vendor or firmware developed by both an in-house designer and sourced from a third-party vendor may also have malicious code. The type of HT or firmware Trojan attack depends on the application. The weakness caused by hardware or firmware Trojan is exploited by an adversary to leak information or to perform DoS attacks. The security framework the ODM should adopt to detect the HT during both pre-silicon and post-silicon validation is discussed in the research literature [44, 45]. It is difficult to device a detection technique which can detect all types of HT. Even in the presence of a security framework to detect HT, the presence of few stealthy HT in the chip is not overruled. Due to these, chips used in critical applications must have run-time defense mechanisms in place to detect the malicious activity during run-time and to protect the crucial assets of SoC from HT, firmware Trojan, and any other kind of DoS attacks by an adversary. Run-time monitoring techniques proposed in [46, 47] are custom-designed for specific processors, and it is suitable for small processors. The volume of data monitored and complexity of the detection module increase for complex processors.

The source of malicious activity can be HT, FT, or from any other type of adverse activity. Acting on malicious activity is wiser during run-time, and such research endeavors can also use hardware performance counters, side-channel analysis, and debug instruments to detect malicious activity.

Hardware Performance Counter (HPC)-Based Malicious Activity Detection HPCs are performance-monitoring modules, which can be used to detect malicious activity also. HPCs collect the information related to event timing, branch predictions, cache misses, etc. during program execution. HPCs can be used in different ways to detect malicious activity as described in the following.

The signature-based technique called SIGDROP proposed in [48] focuses on detecting return-oriented programming (ROP) attacks such as high-level misprediction by return address stack (RAS). ConFirm [49] detects the malicious activity in a firmware with the help of HPCs. ConFirm is targeted to detect malicious activity in embedded systems in which the same firmware code executes at normal circumstances, and any change introduced by an attack is detected using HPCs. Machine learning techniques are applied on HPC data to detect malicious activity. Generally, training of HPC data is performed offline, and the classification engine required to detect malicious activity is developed and deployed. HPC Mall Hunter proposed in [50] monitors the HPC data to detect the malicious behavior. The known data is used for offline training, and database for malicious activity is stored as the matrix and used for attack detection. Reference [51] uses half-space trees to detect the malicious activity of hardware Trojans that introduce abnormal patterns in the data obtained from HPC. This technique is useful in detecting the activation of hardware Trojans that cause DoS and change in functionality of a microprocessor.

Most of the HPC-based malicious activity detection methods depend on complex machine learning techniques and training processes which are performed offline. HPC-based techniques are memory-intensive, and offline training is very slow. Offline training requires large data for training and validation. Because of these factors, HPC-based techniques are not suitable for real-time run-time detection of malicious activity. All sensitive modules or IP cores in the SoC device must be attached with the HPC module, for detection and coordination between the different modules is not easy to implement.

Detecting Malicious Activity Using Side-Channel Analysis Park et al. [52] describe the power side-channel-based instruction-level disassembler which can be used for code tracing, reverse engineering, and integrity checking of firmware/software running on a device. Similar to [52], the work presented in [53] uses power traces extracted through side-channel analysis with machine learning for software integrity verification. Disassembly-based side-channel attacks are more suitable for low-end microcontrollers (8/16-bit), and it is difficult to measure accurate power traces for advanced processors with pipelining, etc. Most of the side-channel analysis schemes need large-scale data collection and processing which make these techniques not attractive for real-time monitoring.

E-IIPS (Enhanced Infrastructure IP for Security) proposed in [54] provides run-time defense against malicious activity. E- IIPS leverages the IEEE 1500 core test standard wrapper and debugs structures (similar to ARM CoreSight) to detect anomalous events on the core I/O. Centralized dedicated processors take defensive measures when unexpected events are detected.

IoT edge nodes should have mechanisms deployed to mitigate the different kinds of malicious activities, irrespective of their source like HT, firmware Trojans, unauthorized software, and DoS-type attacks from the external environment. A programmable centralized facility in the SoC which protects its crucial assets and facilitates the IoT edge node designers to deploy their security requirements through firmware is necessary. The SoC-level defense mechanisms can be used to defend the IoT edge node against not only HT but also from firmware Trojans and other DoS attacks.

In this section, we discussed counterfeiting, debug security, and malicious activity aspects for both SoC security and IoT edge node security. A centralized infrastructure IP is an attractive solution to address the security of both SoC and edge node. In the next section, we present the survey of existing security infrastructure IP from both industry and academia.

4 Centralized IP for SoC Security and IoT Edge Node Security

The modern-day SoC devices consist of specialized infrastructure IP cores for many applications, which are not part of SoC functionality requirement. Infrastructure IP cores are critical to modern-day SoC devices and very useful in chip testing, diagnosis, repair, malfunction detection, etc. [55]. Infrastructure IP for Security (IIPS) described in [56] addresses the issues like counterfeiting, countermeasures against scan-based SCA, and run-time hardware Trojan detection. Inclusion of PUF in IIPS addresses the problem of identification/authentication of the chip, and by securing the IEEE 1500 wrapper interface, scan-based SCA is prevented. In IIPS, path delay-based hardware Trojan detection scheme is also included, but defense against hardware Trojan becomes more important than hardware Trojan detection during run-time. E- IIPS (Extended-IIPS) is an extension to IIPS which includes run-time hardware Trojan defense mechanism, which is described in [54]. E-IIPS is a microprocessor-based module added to basic IIPS to implement the security policies [54, 57]. Firmware of E-IIPS can be upgraded to implement new security policies or any modifications during run-time. E-IIPS supports all features of IIPS and implements security policies as a defense mechanism against malicious activity.

Industry also seriously perceived the idea of centralized IP for security, and there are several such IPs like ARM TrustZone, Intel SGX, Maxim DeepCover, etc.

Centralized SoC security modules also known as trusted platform modules (TPM) like ARM TrustZone [58] and Intel SGX [59] are already adapted by semiconductor vendors which are helpful in edge node identification also offer secure environment for the execution of sensitive processes. TPMs also provide security against physical attacks like tampering and probing, but security issues connected with malicious activity (HT, FT etc.) and debug security are not addressed. The secure firmware update and IoT edge node identification with the existing TPMs like ARM TrustZone store the keys used for encryption in nonvolatile memory, which makes TPMs vulnerable. Rambus has come up with a programmable IP core for security called CryptoManager Root of Trust (CMRT) RT360 [60] which includes 32-bit RISC CPU, AES core, SHA-2 hash core, asymmetric public key engine, secure power management core, TRNG, and key derivation core (KDC). CMRT supports secure boot, secure management of keys, secure upgrade, secure debug, device authentication, and key provisioning. Encrypted key is stored in the nonvolatile memory of the device [60].

The SoC/microcontroller is an important component of IoT edge node and must have some core capabilities to support security needs of an edge node in the form of centralized security infrastructure IP. The necessary core capabilities (list of features in Table II) of such centralized security IP are discussed as follows:

Secure Firmware/Software Management and Update It is implicitly assumed that firmware preloaded on SoC/microcontroller is secure. In the IoT era, the firmware/software updates are performed over the air (OTA) or online. The probability of adversary deploying the malicious firmware is not ruled out. The malicious firmware may compromise the security and privacy of the edge node. Another possibility is a counterfeit/cloned edge node getting updated by an authorized firmware may convert the counterfeit node into a legitimate node. All these dangers must be addressed by a secure update section of the centralized infrastructure. The three firmware security properties are

- System Boot: Secure boot process requires a bootloader stored in secure memory to validate and load the firmware and software. Keys used to authenticate the firmware and software must be stored in secure memory. The entry points like network interfaces and JTAG ports of the system must be inactive during the boot process. Bootloader program used in the boot process must be tied to unique device identifier which helps in thwarting the firmware cloning attack [63].
- Firmware Update: Firmware updates are performed online, and the possibility of illegitimate firmware getting deployed on IoT edge node is not ruled out. The firmware update process and firmware must be secured. The firmware/software must be authenticated cryptographically, and deployment on device should be according to the Transport Layer Security (TLS) standard. The firmware/software must then be validated using the keys stored in secure memory [63, 67].

- Run-time Check: An adversary can modify the firmware during run-time. Thwarting run-time tampering attacks demand countermeasures like run-time integrity check, etc. [63].

Generally, bootloader provided by the SoC vendor and corresponding keys are stored in secure memory and not vulnerable to attacks. Based on this assumption, secure firmware management is designed. Firmware stored in secure memory alone is not sufficient to ensure security, and SoC vendors should leverage hardware security primitives like TRNG or PUF to generate keys for encryption, authentication of firmware, identification of edge node, etc. An endeavor toward designing a holistic secure firmware update process with the abovementioned properties using hardware security primitives is an open research problem. The secure update literature available for few FPGA devices [64] does not reveal more information to discuss their prospects and consequences.

Comprehensive Debug Security Test and debug modules in SoC increase the observability and controllability of the design which is an asset for testing, but it is a threat to security. There are several successful attacks which use test and debug structures to leak secret keys used in ciphers and other trade-related information like DRM keys, etc. [65, 66]. The legacy technique used to protect sensitive chips from attacks is by blowing the fuses during production testing which disconnects the test and debug interfaces from the external world. In modern SoC devices, it is not recommended to permanently disable debug interfaces [68]. Several debug instruments are required at different points of the chip life cycle. So, as a security measure, access control and authentication requirements should be imposed at debug interfaces and entry points. The multilevel access control and authentication requirements are must in test mode when debug instruments are active. JTAG (1149.1) is a common interface used for test and debug, and there are several standards for test and debug like IEEE 1500 and IEEE P1687 (IJTAG). IEEE 1500 is a core test standard to streamline the SoC designs built on IP cores. And the IEEE P1687 standard is used to streamline the access to on-chip debug instruments. These test standards give unprecedented flexibility and easy access to on-chip instruments and test structures which are a big threat to sensitive assets of SoC and cryptographic cores. A sound understanding of the system and access path to sensitive assets is required to place the access control mechanism at appropriate locations [68]. JTAG is used as a primary access to both IEEE 1500 and IJTAG networks. Individual IPs may snoop on system bus or other IP cores. The access control schemes should also address data transactions between the different test networks and IPs along with the external world. The future research should address designing secure test access mechanisms, segregation of on- chip instruments based on their applications, placement of access control mechanism, and securing the reconfigurable scan networks like 1500 and IJTAG in a comprehensive way.

Anti-Counterfeiting/Device Attestation Anticounterfeiting solution is a requirement of a SoC design house to prevent revenue and reputation loss due to counterfeit components. Device attestation is a mandatory requirement in the current day

IoT ecosystem. The IoT edge node security framework should support three categories of attestations: static attestation, run-time attestation, and software attestation [69].

Static Attestation Static attestation is part of the secure boot process in which integrity of the firmware and initial setup is verified. Static attestation also includes the secure update process. Creating infrastructure for static attestation is the primary responsibility of the SoC design house/ODM by incorporating TRNG or PUF, secure memory, and firmware to perform this attestation.

Run-Time Attestation After booting, run-time validation is equally important to detect the malware injection or side-channel attack on the edge device through ports and interfaces. Edge device in the custody of an adversary may face tampering attacks. Malware injection, side-channel attacks, and DoS attacks can be detected through run-time validation and integrity check.

Software Attestation Software attestation is a general software validation procedure followed in personal computers, and mobile phone is followed in IoT edge node at a smaller scale. The security framework designed based on the root of trust coming from hardware sign the software package and assure the integrity and authenticity of the source of the software package.

There are several anticounterfeiting solutions available, and most discussed are split manufacturing, hardware metering, and secure split test (SST) [19]. The hardware metering and SST schemes use PUF or TRNG. So, SST is a suitable technique which can be scalable to address the IoT edge node security. Design and development of a comprehensive scheme which addresses all three categories of attestation using hardware is difficult. Static attestation can be achieved using hardware security primitives like PUF or TRNG, and other two attestations can be designed using a right mix of hardware and software. Hardware security primitives (PUF or TRNG) can be used for key generation in all attestations, which makes the system more secure as the root of trust comes from hardware.

Provisioning Keys for Complete Supply Chain Most of the SoC devices carry sensitive assets like DRM keys, de- featuring bits, configuration fuses, premium content, and secret keys used in cryptographic cores. The provisioning key to protect the IP rights of every stakeholder in the supply chain is a challenging problem. SoC design house must develop countermeasures to prevent counterfeits coming out from untrusted foundries and OSAT centers by developing a technique which generates separate keys for testing at the third-party facility and separate keys for end user. The SoC design house will enable the features in the device through the firmware which will be stored in secure memory. The independent third-party IP core vendor must rely on the SoC design house/integrator to monetize their IP cores [66]. IP core vendors can defend against IP infringements and copyright violations through passive watermarks. Any active metering techniques using cryptography or PUF needs trusted support of the SoC integrator [66]. There are several techniques proposed for key provisioning across supply chains like SST, hardware locking, etc., which can be scaled up and used for key provisioning with the help of firmware. The data transactions between the IP core vendor, SoC integrator (ODM), OEM (original equipment manufacturer), and end user of the gadget must be analyzed to address the security issues of every stakeholder in the supply chain. The future research must address toward standardizing how IP core vendors, SoC integrators, OEMs, and end user can protect their IP, trade secrets, and privacy.

Defense Against HT, DoS, and Other Attacks Run-time defense against HT is a least researched area. Most of the research on HT focuses on HT detection than defense. There are other kinds of attacks on edge nodes like DoS coming through the network, malicious firmware injection and fault injection, which will have results of similar effects. The most vulnerable modules of SoC to these attacks are processors, cryptographic cores, and protocol interfaces. We have discussed different run-time malicious activity detection techniques in Sect. 3. The crucial assets of the SoC-like processor, secure memory locations, cryptographic cores, etc. must be guarded. The continuous monitoring of activity on crucial assets and taking countermeasures against any malicious activity at the SoC level are complex issues. A third-party IP core sourced from an untrusted vendor may snoop into other IP cores and leak confidential information. E-IIPS is an example for a holistic SoC-level HT defense mechanism, in which activities of all IP cores are monitored for malicious activity and reported to centralized IP in which a dedicated security processor running a security policy will take appropriate actions [54]. All IP cores in the SoC are monitored in E-IIPS, and it is not essential in all the cases. Monitoring selected sensitive IP cores through security monitors and centralized facility taking action against malicious activity similar to E-IIPS are attractive solutions against unintended activity.

Countermeasures are designed focusing on the source of the attack than the effects of a successful attack. The generalized intentions of these attacks (HT, malicious code injection, and fault attack) are leaking information and DoS. The future research should focus on developing defense against the effects of attacks, irrespective of the type of attack or source of attack. There is a need to develop methods to monitor the vulnerable situations in sensitive cores and centralized facility to take appropriate decisions when malicious events are detected in different IP cores of the SoC device. This can be achieved by a centralized facility with a security processor which monitors the malicious activity and take security-related decisions.

Trusted Computing Base Better security can be achieved by creating secure execution environments. One way of implementing is through Trusted Execution Environments (TEEs). TEE is defined as a secure area of processor which provides isolated safe execution of security software. All security enforcements are implemented in software stacks that contain boundaries, and such boundaries can be created with the help of hardware through TEE. The basic TEE is achieved through dual execution environment approaches such as ARM TrustZone [58] and TLR [70]. TEE can be implemented in different ways, and the widely used scheme is based on the Dual-Execution Environment (Dual-EE) approach. Dual-EE partitions the system into two execution environments: General Purpose Execution Environment (GPEE) and Secure Execution Environment (SEE) [71]. GPEE runs the general system software, while SEE runs the selected sensitive applications designed to run on SEE. The dual-EE scheme can be implemented in software or hardware or with the right mix of both and must ensure

- Data Separation: Data within one environment cannot read or modify other environment or partition.
- Information Flow Control: Default data transactions between environments cannot occur until they are explicitly permitted.
- Sanitization: There is no information leakage through shared resources.

ARM TrustZone is the most deployed security extension in practice. Processors with TrustZone extensions provide virtualization which enables two processors with two zones: secure zone and normal zone. Isolation is implemented by using hardware controllers, configuration bits, and a new execution mode called monitor mode [58]. ARM TrustZone supports virtual memory, I/O, and interrupt virtualization. This virtualization enables the processor to provide two virtual cores: secure and nonsecure. TrustZone architecture is designed and optimized to implement the dual-EE approach. TrustZone provides protected execution and enables the separate cache entries from both secure and nonsecure execution environments. So, clearing the cache memory during context switching is not necessary. Transition from the secure mode to nonsecure mode or vice versa must go through the monitor mode. With these features, ARM TrustZone provides balanced dual execution environments with all the required features. But TrustZone is bulky for low-end microcontrollers in terms of area and power. So, any trusted execution environment tax the performance of the whole system to a reasonable extent, and it is always not necessary that the SoC devices used in edge node should have a trusted computing base. SoC devices must support hardware acceleration to encryption algorithms, secure, and protected memory which can be accessed only in a privilege mode, which is helpful in the IoT edge node.

Comparative Analysis of Existing Centralized Security IP The comparative analysis of existing centralized security infrastructure IP is presented in Table 2. In the group of commercial TPMs, TrustZone [58] is a widely researched and deployed security platform which provides the hardware-enforced logical separation between secure and normal execution. TrustZone also supports password-based restricted access to debug instruments. ARM TrustZone supports most of the security requirements of both SoC ODMs and IoT edge node security, but TrustZone does not fit into low-power/low-end (8/16-bit) microcontrollers/SoC devices [61]. Most of the commercial TPMs except Maxim DeepCover [62] suffers from larger footprint and high power consumption. IoT edge nodes use low-end microcontrollers to collect data (e.g., sensor nodes, etc.), and complex TPMs are not suitable for microcontrollers used in lightweight IoT devices. Maxim DeepCover [62] which is suitable for low-end microcontrollers does not support debug security and defense against

Table 2 Comparative analysis of hardware security and IoT edge device security

Security technique	Name	Secure enclave	Device attestation/ anti counterfeiting	Provisioning keys for wafer sort, final test, and DRM	Comprehensive debug security	Defense against HT, DoS and software attacks	Secure firmware/ software update	Remarks
Commercial TPM	ARM TrustZone [58]	Supported	Supported. Device key stored in device can be used for attestation through software. Static device ID	Software-based DRM support.	Supported. Access control-based protection to processor and system debug components	Software attacks defense supported. DoS protection partially supported	Supported. RSA is used for encryption of binaries	Power hungry and high footprint. Not suitable for microcontrollers used in IoT edge nodes
	Intel SGX [59]	Supported	Software-based attestation	Software-based DRM support.	Partially supported	Defense against DoS attacks supported	Not supported	Power hungry and high footprint
	Maxim DeepCover [62]	Not supported	Dynamic device ID based on PUF circuit	Software-based DRM support.	Not supported.	Not supported	Supported	Suitable for low-end SoC. Lacks Debug security.
	Rambus CryptoManager [60]	Supported	Supported	Supported	Supported	Supported	Supported	May not fit into Low-end devices. Details not available
Infrastructure IP for security	IIPS [56]	Not supported	Supported	Not	Partially supported	Not supported	Not supported	Secure update and full debug security not supported.
	E-IIPS [54]	Not supported	Supported	supported	Partially supported	Supported	Not supported	

several types of attacks. Rambus CryptoManager is a more holistic TPM which supports all required security features. Rambus CryptoManager (CMRT) is built around the RISC-V processor which is designed to securely run the sensitive processes and support full suite of security services like secure boot, software integrity checking, remote attestation, secure firmware upgrade, DRM for sensitive assets, device authentication, secure debug, cryptographic hardware accelerators, and key provisioning to protect the IP rights of all stakeholders in the supply chain from the SoC vendor to end user. CMRT whitepaper [60] claims that it provides protection against a wide range of side-channel attacks and fault injection which are not explained [60]. Apart from commercial TPMs, we can also find infrastructure IPs for security proposed in academic research.

Infrastructure IP for Security (IIPS) addresses three major security features: protection against scan-based SCA, PUF-based device authentication, and run-time HT detection. The major limitation of IIPS is that it does not support run-time defense against HT or malicious activity. HT detection is a very useful activity before fabrication of the chip. Run-time defense against several attacks including HT will be crucial once the chip is produced. E-IIPS (Extended IIPS) is an extension to IIPS in which run-time defense is addressed in a more comprehensive way, and it is assumed that E-IIPS supports all features of IIPS and also supports run-time defense against malicious activity. E-IIPS defines a holistic system-level security policy in which security wrapper is integrated with each constituent IP core of SoC. Security wrapper is an extension to the IEEE 1500 core test standard and ARM CoreSight IP interface standard. Security wrappers detect the local events connected with security and communicate with the centralized E-IIPS module. E-IIPS will take suitable steps in the case of security breach to mitigate the effect of attack. Security policy is programmed as firmware into E-IIPS. This makes E-IIPS scalable such that new security policies can be updated in firmware to address the new security threats. E-IIPS is a holistic and scalable solution, but it comes with few limitations like area overhead and power consumption and does not address the security of E-IIPS itself, when an adversary deploys an illegitimate firmware which completely defeats the purpose of centralized security IP. E-IIPS does not address important security requirements like secure firmware update and on-chip instrumentation security.

A review on centralized infrastructure which addresses the security requirements of both SoC vendors and IoT edge nodes is presented. Detailed analysis on Rambus CMRT is not possible as it is a proprietary technology, and all other commercial TPM-based centralized schemes are discussed in this chapter. Centralized Security IP is the need of the day, and serious research is going on both in industry and academia to design different types of security IPs to meet the requirements of various applications. The generalized features the future centralized security infrastructure IPs must support are as follows:

- Secure boot, secure firmware/software update, and run-time integrity checking.
- Device authentication and identification.
- Hardware accelerators for cryptography.
- Comprehensive test and debug security.

- Defense against different types of SCA.
- Provisioning keys for all stakeholders in supply chain from the SoC vendor to end user.
- Must be programmable to update the firmware for future threats.
- Secure storage of keys and other sensitive assets.
- Trusted computing base.

5 Conceptual Foundations to SIIP Architecture

In this chapter, we propose a conceptual foundation for Security Infrastructure IP (SIIP), which supports most of the security requirements of both IoT edge nodes and SoC vendors. The conceptual block diagram of Security Infrastructure IP (SIIP) is shown in Figs. 2 and 3. The design principles followed to achieve the necessary security features of centralized IP to support both SoC security and IoT edge node security are as follows:

- Addressing Debug Security: Debug security can be achieved by implementing password-based access restriction and authentication schemes at debug interfaces. The three major test protocols used in modern SoC are conventional JTAG,

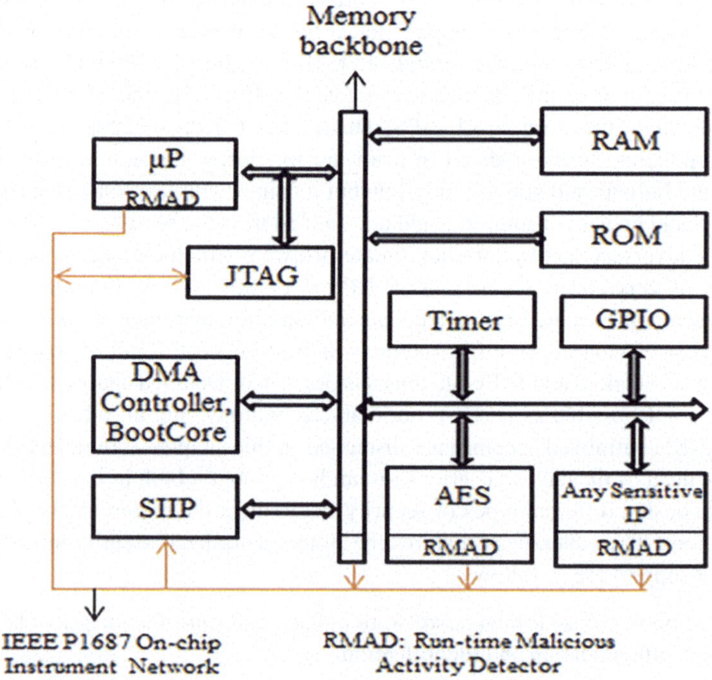

Fig. 2 SoC device with SIIP

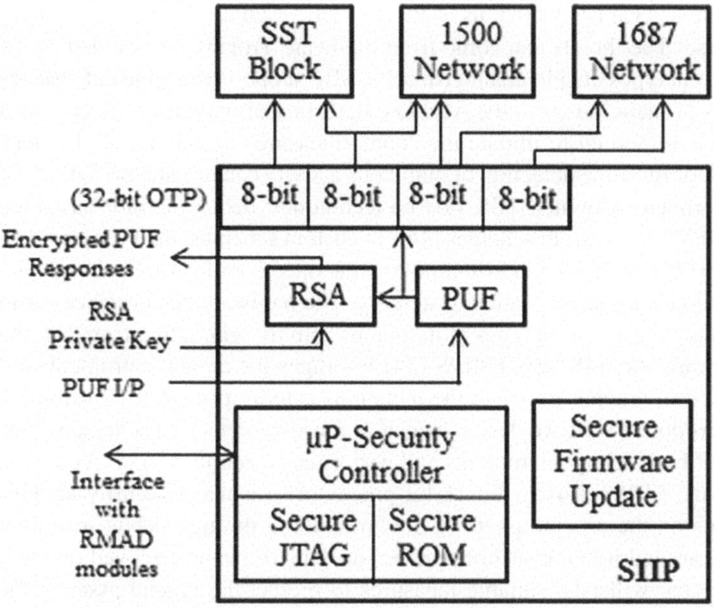

Fig. 3 Architecture of conceptual SIIP

IEEE 1500 core test standard for core testing, and IJTAG network for streamlined access for on-chip instruments. The placement of authentication mechanism should not modify the test standards. The interface logic between 1500 and external world is generally achieved by connecting to conventional 1149.1 JTAG. The glue logic between two test standards (1149.1 and 1500) can be as small as two OR gates to complex structures like a processor. The authentication mechanism can be part of an interface glue logic. The bridge between different test standards is not standardized, and these structures can be modified to incorporate an authentication mechanism. PUF is used to generate the keys for authentication. A security framework can be designed by incorporating the access restriction and authentication mechanisms in glue logic which acts like a bridge between different test protocols.

- Addressing Counterfeiting/Device Attestation and Provisioning Keys for IP Protection: Secure Split Test (SST) is a more comprehensive solution than other anticounterfeiting solutions like hardware metering [19], split manufacturing [72], and hardware locking [73]. The major advantage of the SST technique is that it prevents counterfeit chips coming out from both the untrusted foundry and test centers. SST can be scalable to address the problem of counterfeit IoT nodes and provisioning keys for IP protection [74].
- Secure Firmware Update: Most of the available solutions to this problem are described in industry whitepapers. The secure firmware update scheme described in [75] is adopted with suitable modifications in the proposed SIIP.

- Defense Against Malicious Activity: The sources of malicious activity are diverse. The threats can come from hardware Trojans, malicious firmware, and different types of side-channel attacks. SIIP supports run-time defense against all types of malicious activity. Accurate detection of malicious activity during run-time is important to initiate the countermeasures to mitigate it. The techniques used for run-time detection of malicious activity can be categorized as: Hardware Performance Counter (HPC)-based techniques; debug monitor-based technique (E-IIPS), and custom schemes [46]. In custom schemes, the Processor Protection Unit (PPU) proposed in [46] monitors the opcode of instructions and other internal signals to detect malicious activity. The hardware performance counters are used to detect several types of malicious activity originating from HT, malicious firmware, etc. [48, 49]. E-IIPS [54] leverages the debug instruments and IEEE 1500 test wrapper to detect the malicious activity primarily focusing on HT. In the proposed SIIP, we design a custom run-time detection scheme which makes use of both debug instruments and performance counters to detect the malicious activity. SIIP consists of a 16-bit processor to manage security activities. The events on the crucial assets of SoC monitored through debug instruments and HPC are fed into the security processor. The firmware deployed on the security processor will take suitable measures to protect the crucial assets of SoC and mitigate the malicious activity.
- Trusted Computing Base: Trusted computing is required for highly data-intensive applications, and it is uneconomical to have full-fledged trusted computing base in IoT edge nodes due to constraints like power consumption, etc. Trusted computing is highly useful in processors used in higher hierarchies of IoT ecosystem like data accumulation and data crunching [76]. In the proposed architecture, SIIP targeted to 8/16-bit microcontrollers used in IoT edge nodes does not require a trusted computing base. SIIP does not support a separate secure execution environment.

The objective of SIIP is to provide a scalable security infrastructure IP for SoC designs to protect against security threats. The idea of the proposed architecture provides a comprehensive approach to SoC security which is helpful for SoC manufacturers, IoT edge node OEMs, and end-user. We are working on SIIP architecture, and we have discussed the necessity of such a centralized security IP and its features. The representative diagram of SoC architecture with SIIP is shown in Fig. 2. RMAD (Runtime Malicious Activity Detector) attached to sensitive IP cores like processors, cryptographic cores, etc. is connected to SIIP through the IJTAG network. SIIP is connected to the system bus to monitor the events. SIIP will take suitable countermeasures against malicious activity in the sensitive cores and system bus. The internal block diagram of SIIP is shown in Fig. 3.

The core components of the SIIP are microcontroller, PUF, RSA, 32-bit OTP memory, and a secure firmware update block. The microcontroller in SIIP includes the secure memory and secure JTAG. Only ODM (SoC design house) will have an access to use SIIP microcontroller to load the security policy firmware into secure ROM. Secure JTAG is used for deploying the security firmware updates, and its access is restricted to the SoC design house. The security firmware monitors the events on sensitive cores using RMAD module and takes suitable measures against malicious activity. The Physical Unclonable Function (PUF) is used to generate keys for all security purposes. The key generated by PUF is stored in OTP (32-bit) memory and used in the security design of different networks like IEEE 1500 and IJTAG. The same key can be used in the secure firmware update section also. The PUF section includes the RSA encryption module for encrypting the PUF responses. The PUF CRP collection is performed by the SoC manufacturer for hardware metering [25]. The probability of successful machine learning attacks on PUF designs is high when the adversary gets an access to sufficient number of CRP pairs to use in the machine learning process [77]. It is necessary to secure the CRP pairs as most of the CRP collection is performed during production testing in OSAT centers. The response from the PUF is encrypted using the RSA encryption block, and public key for the encryption and design house will share the public key with the OSAT center. In SIIP, only the microcontroller section is restricted from user applications. The PUF section is available for user applications, and PUF can be used for authentication and identification purposes. The secure firmware update section consisting of cryptographic cores also uses the 32-bit key stored in OTP memory to facilitate the safe firmware/software update. The design and implementation of SIIP is under progress, and important issues which require attention are: Test Access Mechanism (TAM), handling activation of security mechanisms before and after testing, and final activation of security configuration of the chip before shipping it to the customer. The SIIP implementation work is under progress, and we expect SIIP to provide a systematic approach to implement SoC security and IoT edge node security.

6 Conclusions

This survey attempted to summarize the SoC security and IoT edge node security issues and countermeasures against them. We have discussed the problems, challenges, and available solutions for both SoC security and IoT edge node security. The objective of this chapter was to give the reader an opportunity to understand how most of the SoC security and edge node security issues are deeply connected and solutions designed for SoC security can be directly used or can be scaled up to address the issues of IoT edge node security.

Second, it is difficult to address the IoT edge node security holistically without a suitable SoC/microcontroller with security properties. SoC vendors must incorporate the security features with IoT edge node security to make their chips attractive to IoT edge node developers.

Third, this survey attempts to analyze the reasons why researchers in industry and academics are working toward more comprehensive solutions to address security issues in the form of centralized security infrastructure IP. Finally, we present the security features the centralized security infrastructure IP should possess and a conceptual foundation to design centralized security IP as a comprehensive solution.

Given the wide applicability of centralized IP, both academics and industry should proactively involve in research which can successfully address most of the SoC and IoT edge node security issues. Centralized IP described serves the security requirements of IoT edge node comprehensively for sensor network applications.

References

1. CISCO. (2014). *The internet of things reference model.* [Online]. Available: http://cdn.iotwf.com/resources/71/IoT_Reference_Model_White_Paper_June_4_2014.pdf
2. Mosenia, A., & Jha, N. K. (2017, October–December 1). A comprehensive study of security of internet of things. *IEEE Transactions on Emerging Topics in Computing, 5*(4), 586–602.
3. Hunt, G., Letey, G., & Nightingale, E. B. *The seven properties of highly secure devices.* Microsoft Research NExT Operating Systems Technologies Group. https://www.microsoft.com/en-us/research/wp-content/uploads/2017/03/
4. Alrawais, A., Alhothaily, A., Hu, C., & Cheng, X. (2017, March). Fog computing for the internet of things: Security and privacy issues. *IEEE Journal of Internet Computing, 21*(2), 34–42.
5. Konstantinou, C., et al. (2016). *Taxonomy of firmware trojans in smart grid devices* (pp. 1–5). IEEE 2016 IEEE Power and Energy Society General Meeting (PESGM).
6. Tehranipoor, M., & Koushanfar, F. (2010, January/February). A survey of hardware Trojan taxonomy and detection. *IEEE Design and Test of Computers, 27*(1), 10–25.
7. Zhou, Y., & Feng, D. (2005). Side-Channel attacks: Ten years after its publication and the impacts on cryptographic module security testing. *IACR Cryptology ePrint Archive, 2005*, 1–388.
8. Mao, B., et al. (2018, September). Quantitative analysis of timing channel security in cryptographic hardware design. *IEEE Transactions on Computer- Aided Design of Integrated Circuits and Systems, 37*(9), 1719.
9. Da Rolt, J., et al. (2014, March). Test versus security: Past and present. *IEEE Transactions on Emerging Topics in Computing, 2*(1), 50.
10. Miura, N. (2015, June). EM attack sensor: Concept, circuit, and design- automation methodology. 52nd ACM/EDAC/IEEE Design Automation Conference (DAC).
11. Barenghi, A., Breveglieri, L., Koren, I., & Naccache, D. (2012, November). Fault injection attacks on cryptographic devices: Theory practice and countermeasures. *Proceedings of the IEEE, 100*(11), 3056–3076.
12. Ali, S., et al. (2011). Multi-level attacks: An emerging security concern for cryptographic hardware. *DATE, 2011*, 1–4.
13. Martin, T., Hsiao, M., Ha, D., & Krishnaswami, J. (2004). Denial-of-service attacks on battery-powered mobile computers. In *Proceedings second IEEE annual conference on pervasive computing and communications* (pp. 309–318). IEEE Computer Society.
14. Cardenas, A. A., et al. (2011). Attacks against process control systems: Risk assessment, detection, and response. In *Proceedings of the 6th ACM conference on computer and communications security* (pp. 355–366).
15. Parno, B., et al. (2005). Distributed detection of node replication attacks in sensor networks. In *Proceedings IEEE symposium on security and privacy* (pp. 49–63).
16. Juels, A., Rivest, R. L., & Szydlo, M. (2003). The blocker tag: Selective blocking of RFID tags for consumer privacy. In *Proceedings of the 10th ACM conference on computer and communications security* (pp. 103–111).

17. Rostami, M., Koushanfar, F., & Karri, R. (2014, August). A primer on hardware security: Models, methods, and metrics. *Proceedings of the IEEE, 102*(8), 1283–1295.
18. Cyient White paper. The-definitive-guide-to-outsourcing-semiconductor- design-projects-what-you-need-to-know-now. www.cyient.com.
19. Kumar, K. S., et al. (2016, September). Secure split techniques to prevent IC piracy for IoT devices. *Elsevier-Journal of VLSI Integration*. http://www.sciencedirect.com/science/article/pii/S0167926016300657.
20. Kumar, K. S., et al. (2015). An improved AES hardware Trojan benchmark to validate Trojan detection schemes in an ASIC design flow. *IEEE VLSI Design and Test (VDAT)*.
21. Tehranipoor, M., et al. (2015). *Counterfeit integrated circuits: Detection and avoidance*. Springer. ISBN: 3319118234 9783319118239.
22. Sudeendra Kumar, K., et al. (2016). *Securing IEEE 1687 standard on-chip instrumentation access using PUF*. IEEE International Symposium on Nanoelectronic and Information Systems (iNIS).
23. Saha, D., & Sur-Kolay, S. (2011). SoC: A real platform for IP reuse IP infringement and IP protection. *VLSI Design, 2011, January*, 1.
24. Sengupta, A., et al. (2018). *Reusable intellectual property core protection for both buyer and seller*. IEEE International Conference on Consumer Electronics (ICCE).
25. Koushanfar, F., & Qu, G. (2001). *Hardware metering* (pp. 490–493). Proceedings of the ACM/IEEE Design Automation Conference.
26. Samsung. Samsung KNOX. www.samsungknox.com.
27. Wang, J., et al. (2018, January). Enabling security-enhanced attestation with intel SGX for remote terminal and IoT. *IEEE Transactions on Computer-Aided Design of Integrated Circuits and Systems, 37*(1), 88.
28. Zeitouni, S., et al. (2017). *ATRIUM: Runtime attestation resilient under memory attacks*. IEEE/ACM International Conference on Computer-Aided Design (ICCAD).
29. Yang, K., et al. (2018, April). ReSC: An RFID-enabled solution for defending IoT supply chain. *ACM Transactions on Design Automation of Electronic Systems (TODAES), 23*(3), 29.
30. Johnson, A. P., et al. (2015, April-June 1). A PUF-Enabled secure architecture for FPGA-based IoT applications. *IEEE Transactions on Multi-Scale Computing Systems, 1*(2).
31. Chatterjee, U. (2017, July). A PUF-based secure communication protocol for IoT. *ACM Transactions on Embedded Computing Systems (TECS) – Special Issue on Embedded Computing for IoT, Special Issue on Big Data and Regular Papers, 16*(3), Article No. 67.
32. DeepCover Embedded Security Technology, Maxim Integrated, https://www.maximintegrated.com/en/products/embedded-security/deepcover.html
33. Secure Device Manager for Intel® Stratix® 10 Devices Provides FPGA and SoC Security, https://www.intel.com/content/dam/www/programmable/us/en/pdfs/literature/wp/wp-01252-secure-device-manager-for-fpga-soc-security.pdf
34. https://www.intrinsic-id.com/
35. Developing Tamper-Resistant Designs with Zynq UltraScale+ Devices, https://www.xilinx.com/support/documentation/application_notes/xapp1323-zynq-usp-tamper-resistant-designs.pdf
36. Stollon, N. (2010). *On-chip instrumentation: Design and debug for systems on chip*. Springer.
37. Yang, B. et al, (2004). Scan based side channel attack on dedicated hardware implementations of data encryption standard. Proceedings of international test conference (ITC).
38. IEEE Computer Society. IEEE standard testability method for embedded core-based integrated circuits, IEEE Std. 1500–2005.
39. IJTAG. IJTAG – IEEE P1687. http://grouper.iee.org/groups/1687.
40. Lee, K. (2016). A brief review on JTAG security. 10th international conference on innovative mobile and internet services in ubiquitous computing.
41. Kumar, S., et al. (2016). Securing IEEE 1687 standard on-chip instrumentation access using PUF. *Proceedings of the IEEE iNIS*, 56–61.

42. Sudeendra Kumar, K., et al. (2018, July). PUF based secure test wrapper for SoC Testing. IEEE annual symposium on VLSI (ISVLSI).
43. Mayhew, M. (2016). An overview of hardware-level statistical power analysis attack countermeasures. *Journal of Cryptographic Engineering, 7*, 213.
44. Sudeendra Kumar, K., et al. (2015). An improved AES Hardware Trojan benchmark to validate Trojan detection schemes in an ASIC design flow. *IEEE VLSI Design and Test (VDAT).*
45. Dupuis, S., et al. (2018, April). Protection against hardware trojans with logictesting: Proposed solutions and challenges ahead. *IEEE Design and Test, 35*(2).
46. Dubeuf, J. et al. (2013, July). Run-time detection of hardware Trojans: The processor protection unit. 18th IEEE European Test Symposium (ETS).
47. Backer, J. et al. (2014, October). Reusing the IEEE 1500 design for test infrastructure for security monitoring of Systems-on-Chip. IEEE international symposium on defect and fault tolerance in VLSI and nanotechnology systems (DFT).
48. Wang, X., & Backer, J., (2016). SIGDROP: Signature-based ROP detection using hardware performance counters. arXiv preprint arXiv:1609.02667.
49. Wang, X. et al, (2015, November). ConFirm: Detecting firmware modifications in embedded systems using hardware performance counters. IEEE/ACM International Conference on Computer-Aided Design (ICCAD).
50. Bahador, M. B., Abadi, M., & Tajoddin, A. (2014). HPCMalHunter: Behavioral malware detection using hardware performance counters and singular value decomposition. In *Proceedings of the ICCKE* (pp. 703–708). IEEE.
51. Elnaggar, R., et al. (2017). *Run-time hardware trojan detection using performance counters.* International Test Conference.
52. Park, J. (2017, July). Using power clues to hack iot devices: The power side channel provides for instruction-level disassembly. *IEEE Consumer Electronics Magazine, 6*(3), 92.
53. Msgna, M., Markantonakis, K., Naccache, D., & Mayes, K. (2014). Verifying software integrity in embedded systems: A side channel approach. In *International Workshop on Constructive Side-Channel Analysis and Secure Design* (pp. 261–280). Springer.
54. Basak, A., et al. (2017). Security assurance for system-on-chip designs with untrusted IPs. *IEEE Transactions on Information Forensics and Security, 12*, 1515–1528.
55. Zorian, Y., et al. (2002, October). *Embedded memory test and repair: Infrastructure IP for SoC yield* (Vol. 20, p. 58). Proceedings of International Test Conference.
56. Wang, X., et al. (2015, August). IIPS: Infrastructure IP for secure SoC design. *IEEE Transactions on Computers, 64*(8).
57. Basak, A. et al., (2015, November). A Flexible architecture for systematic implementation of SoC security policies. IEEE/ACM International Conference on Computer Aided Design (ICCAD).
58. ARM TrustZone Whitepaper, https://developer.arm.com/technologies/trustzone
59. Costan, V. et al, Intel SGX Explained. https://eprint.iacr.org/2016/086.pdf
60. Rambus CryptoManager Root of Trust (CMRT). https://www.rambus.com/blogs/introducing-the-rambus-cryptomanager-root-of-trust-cmrt/
61. Rahman, F., Farmani, M., Tehranipoor, M., & Jin, Y., (2017). Hardware-assisted cybersecurity for IoT devices. IEEE Microprocessor Test, Security, and Verification Conference (MTV).
62. Maxim DeepCover Technology. https://www.maximintegrated.com/en/products/embedded-security/deepcover.html
63. Hardening the security of Azure IoT Edge. https://azure.microsoft.com/en-in/blog/hardening-the-security-of-azure-iot-edge-2/
64. Intel FPGA Whitepaper, Secure device manager for Intel® Stratix® 10 Devices Provides FPGA and SoC Security, https://www.intel.com/content/dam/www/programmable/us/en/pdfs/literature/wp/wp-01252-secure-device-manager-for-fpga-soc-security.pdf
65. Tehranipoor, M., & Lee, J. (2011, August). Protecting IPs against scan-based side-channel attacks. In *Introduction to hardware security and trust.* Springer.

66. Wang, X. et al., (2018). Secure scan and test using obfuscation throughout supply chain. IEEE Transactions on Computer-Aided Design (TCAD).
67. Tehranipoor, M., Guin, U., & Bhunia, S. (2017). Invasion of the hardware snatchers: Fake hardware could open the door to malicious malware and critical failure. *IEEE Spectrum, 54*, 36.
68. Ray, S. et al. (2015). Correctness and Security at odds: Post-silicon validation of modern SoC designs. 52nd ACM/EDAC/IEEE Design Automation Conference (DAC).
69. NXP firms up firmware security. https://blog.nxp.com/security/nxp-firms-up-firmware-security
70. Santos, N. et al. (2011, March). Trusted language runtime (TLR): Enabling trusted applications on smartphones. Proceedings of 12th workshop on mobile computing systems and applications.
71. Maene, P., et al. (2018, March). Hardware-Based Trusted Computing Architectures for Isolation and Attestation. *IEEE Transactions on Computers, 67*, 361.
72. Rajendran, J. J. V., Sinanoglu, O., & Karri, R. (2013). Is split manufacturing secure? In *Proceedings of the conference on design, automation and test in Europe* (pp. 1259–1264).
73. Juretus, K., (2018, May). Time domain sequential locking for increased security. IEEE International Symposium on Circuits and Systems (ISCAS).
74. Zhang, D., et al. (2018, November). An on-chip dynamically obfuscated wrapper for protecting supply chain against IP and IC Piracies. *IEEE Transactions on Very Large Scale Integration (VLSI) Systems, 26*(11), 2456.
75. Guillen, O. et al. (2015, September). Crypto-Bootloader: Secure in-field firmware updates for ultra-low MCUs. www.ti.com.
76. Vavala, B., (2017). Secure tera-scale data crunching with a small TCB. 47th annual IEEE/IFIP international conference on Dependable Systems and Networks (DSN).
77. Delvaux, J., Machine learning attacks on PolyPUFs, OB-PUFs, RPUFs, LHS-PUFs, and PUF–FSMs. https://eprint.iacr.org/2017/1134.pdf.
78. Wang, X., Guo, Y., Ramhan, T., Zhang, D., & Tehranipoor, M. (2017). DOST: Dynamically obfuscated wrapper for split test against IC piracy. In *Asian Hardware Oriented Security and Trust Symposium (AsianHOST)* (Vol. 2017, pp. 1–6).
79. Lee, K.-J., Liu, C.-A., & Wu, C.-C. (2022). A dynamic-key based secure scan architecture for manufacturing and in-field IC testing. *IEEE Transactions on Emerging Topics in Computing, 10*(1), 373–385. https://doi.org/10.1109/TETC.2020.3021820
80. Sudeendra Kumar, K., Rekha, S. S., Koushik, A., Swain, A., & Mahapatra, K. (2023). Lightweight secured split test technique with RMA capability to prevent IC counterfeiting. In *IEEE International Symposium on Smart Electronic Systems (iSES)* (Vol. 2023, pp. 191–196).
81. Calzada, P. E., Sami, M. S. U. I., Azar, K. Z., Rahman, F., Farahmandi, F., & Tehranipoor, M. (2023). Heterogeneous integration supply chain integrity through blockchain and CHSM. *ACM Transactions on Design Automation of Electronic Systems, 29*(1), 25.

Power Analysis Attack Resistant ASIC Implementation of AES Hardware Accelerator Based on Threshold Implementation

P. Saravanan, S. Shanthi Rekha, S. Rajesh Srivatsav, S. Prashanth, M. Aravindh, G. Lakshmanan, R. K. Kirubhas Shankar, and A. R. Abdul Rajak

1 Introduction

The Advanced Encryption Standard (AES) algorithm is a symmetric block cipher algorithm used for secure data transfer worldwide [1]. It has been the industry standard for the encryption of sensitive data and has not been cracked yet. AES, being a symmetric block cipher, uses the same key for both encryption and decryption. Since the algorithm is implemented as software in most of the security-focused systems, the implementation of a dedicated hardware peripheral module would allow the main processor to perform other time-critical operations. Several hardware implementation works of AES were reported in the literature to improve the throughput [2, 3] and reduce the area and power consumption [4].

1.1 Power Analysis Attacks

The software and hardware implementations of cryptographic ciphers have unwanted leakages that can be leveraged to extract the secret key of the device by a type of attack, namely side-channel attack (SCA) [5]. The unwanted leakages can be in the form of power consumption, electromagnetic (EM) emanations, acoustic emanations, time taken for computation, etc. Among the various side channels,

P. Saravanan (✉) · S. Shanthi Rekha · S. Rajesh Srivatsav · S. Prashanth · M. Aravindh · G. Lakshmanan · R. K. Kirubhas Shankar
Department of Electronics and Communication Engineering, PSG College of Technology, Coimbatore, Tamilnadu, India

A. R. Abdul Rajak
Department of Electrical and Electronics Engineering, Birla Institute of Technology and Science, Dubai Campus, United Arab Emirates

© The Author(s), under exclusive license to Springer Nature Switzerland AG 2025
A. Mishra et al. (eds.), *Hardware Security: Challenges and Solutions*,
https://doi.org/10.1007/978-3-031-81213-2_5

power-based side channels have been extensively studied in the literature in the name of power analysis attacks (PAAs) [5], which pose a strong vulnerability to the cryptographic cipher implementation, and it is important to mitigate the same. Also, some works were done to secure the hardware implementation of AES against cryptanalysis [6, 7], electromagnetic attacks [8], fault analysis [9] and power analysis attacks [10–15] using threshold implementation [16] as an efficient countermeasure. Moreover optimization is also done to refine and improve the efficiency of the threshold implementation against power analysis attacks [17–21]. Besides, an asynchronous design [22] is also considered as an effective countermeasure to counteract side-channel attacks and few more techniques such as shuffling and code polymorphism [23] are reported in the literature to improve the security of AES implementations. This work has implemented a power analysis attack-tolerant AES hardware accelerator at a low cost to be used in IoT applications and edge devices. The threshold implementation (TI) technique is a proven countermeasure against differential power analysis (DPA) attacks, and hence the same is chosen in this work.

The major contributions to this chapter are as follows:

- A two-share TI-based AES S-box is derived based on which a power analysis attack-tolerant AES hardware accelerator is designed.
- The ASIC implementation of the AES hardware accelerator is described in detail, along with the results at each stage for the fabricated IC.

2 Threshold Implementation: Countermeasure to PAA

The masking countermeasure [24, 25] aims to suppress the relationship between the power consumption and the data processed by obscuring the actual data with a random number. The effect of the randomness is removed at the end of the process so that the output of the cipher is not disturbed. The input data is split into two or more shares by XORing the input data with the random numbers generated by the pseudo-random number generator (PRNG). The complete data path of the cipher is modified to process the shared data. The masking technique proposed in the literature showed leakages and glitches. Hence, more properties or requirements were added to masking to make it resistant to glitches, and the technique goes by the name TI [26].

The three main requirements of TI are as follows:

1. Correctness: Since the input is split into two or more shares, the XOR of the shares at the output should produce the actual cipher text.
2. Uniformity: The input shares should be uniform in nature, i.e., the shares of the input vector should occur in uniform probability such that at each stage of the data path, the data processed and thereby the outputs are uniformly distributed.
3. Non-completeness: When the data is split into two or more shares, each processing element in the data path should be independent of at least one share.

Figure 1 shows a depiction of the three properties of TI. A TI implementation that satisfies these properties is proven to be resistant against a p-probing model

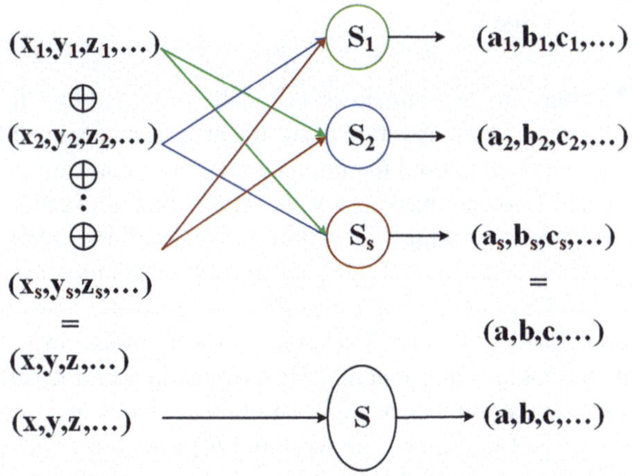

Fig. 1 TI properties

adversary in which the attacker can probe the "p" wire inside the circuit to read the values of the data processed. Correctness is a default property of masking and TI since the functionality verification of the circuit will ensure the same. The noncompleteness property can be satisfied by pipelining the data-path functions at sufficient intervals. This ensures that the computation of a function does not depend on all the shares at the same instant of time. The uniformity property can be satisfied by introducing additional fresh random numbers at the pipelining stages. Hence, TI has the following advantages: (i) The noncompleteness property ensures that the intermediate results do not leak any information about the secret value since each share is independent of at least one input. (ii) Sufficient resistance against PAA is obtained even in the presence of glitches. (iii) The TI technique is realizable with the standard RTL design flow. TI is the only technique that resists the probing adversary model, i.e., an "*s*" shared based circuit is protected against the "*s*-1" probing model. An attacker who can probe and read "*s*-1" wires inside the circuit will not be able to gain sufficient information to retrieve the secret key.

Let "*s*" denote the number of shares of the data, "*d*" denote the algebraic degree of the vulnerable function, which is calculated by the maximum number of inputs of a min-term in its ANF, and "*t*" represent the degree of security. In the earlier works [26–28], the minimum number of shares of any function is represented as.

$$s_{in} = td + 1 \tag{1}$$

In the later work [29], the number of shares was further reduced as below:

$$\text{Input shares}: s_{in} = d + 1 \tag{2}$$

$$\text{Output shares}: s_{out} = (d + 1)^t \tag{3}$$

Based on Eqs. (2) and (3), the TI implementation for AES s-box is derived, and the two-share-based data path is constructed, as explained in subsequent sections. Though the reduction in shares results in a lower area cost, there is an increase in randomness to satisfy uniformity at each pipelining stage.

3 Related Works

The masking technique was originally proposed by Chari et al. [24] and Messerges [25] with a demonstration of sufficient security against SPA and first-order DPA. The later works explored the implementation of masking in two ways, namely arithmetic and Boolean masking for the AES S-box, and efficient conversions between the two. When masking was exposed to leak information through glitches, the TI proposed by Nikova et al. [26] evolved as a technique to defend against the weakness of masking. In general, for a "d^{th}" order security of any arbitrary function with algebraic degree "t", the TI scheme uses a minimum of $(td + 1)$ shares. Based on this methodology, Bilgin et al. [28] have explored three different AES architectures by varying the number of input and output shares and analyzed the performance and security of the same. The work in [29] was to further reduce the number of shares from $(td + 1)$ to $(d + 1)$, with the condition that the input shares should be independent. The work in [30] was to implement the $(d + 1)$ share-based three-stage pipelined AES s-box with a randomness of 12, 24, and 28 bits at each pipelining stage for uniformity. A serial AES architecture was developed based on this s-box, and the performance analysis showed the lowest area when compared with the other works in the literature.

In hardware implementation, the challenges involved in generating the random bits are not explicit, and hence further literature aims to minimize the randomness using various techniques. One such technique was "changing of guards" wherein, except for the first S-box, the randomness for the subsequent S-boxes was derived from the previous input shares [31]. Since the shares are inherently uniform and random in nature, the uniformity of the subsequent S-box outputs is preserved. Thereby, the total randomness required in this technique was that of the randomness required for sharing the inputs initially and those of the first S-box. Wegener and Moradi's work in [32] was to apply the change of guards [31] technique to a four-share AES s-box, the latter implemented using cubic bijections. The AES data path of [32] was based on two shares and the S-box required four shares. The authors state that 24-bit randomness was used once during AES power-up, and the subsequent encryptions used the random numbers stored in the guard registers of the previous encryption. However, they have achieved 10x more latency compared to other similar designs and a higher area occupancy. Another technique was randomness reuse [33], in which, wherever possible, the random bits were reused among the components of the same S-box. The authors have also worked to ensure that the effect of random bits is maintained for the software implementation of AES. In a recent work [34], the author has adapted the reversible logic concept using additional inputs, in which the nonlinear components are made reversible, thereby maintaining uniformity. The additional outputs generated are used as additional inputs (fresh random bits) to the next S-box computation to achieve uniformity.

Apart from the randomness requirement, another observation of the TI and masking techniques is that the implementation of one particular order of security is susceptible to the higher orders of PAA. Though higher order attacks are challenging for an adversary, there is vulnerability.

Based on the works in the literature, the $(d + 1)$-share-based three-stage pipelined S-box concept proposed in [30] is implemented to realize a two-share-based AES architecture.

4 Proposed Architecture

The proposed architecture is explained as follows.

4.1 Two-Share TI Implementation of AES S-Box

This section explains the TI derivation for the AES architecture along with the ANF expressions of the combinational sbox.

For the first-order security $d = 1$, the number of input shares is fixed at 2. While the linear components of the data path can be replicated based on the number of shares, the nonlinear blocks have to be derived explicitly to maintain the correctness of the output. The combinational CFA-based S-box block consists of linear components, namely isomorphic mapping, XOR, squarer, constant multiplier, and inverse isomorphic masking, that can be replicated twice to process the two input shares. However, it is required to derive TI for the nonlinear blocks, namely the Galois field (GF)(2^4) multiplier and inverter of the s-box, using the algebraic normal form (ANF) expressions. Since the multiplier is implemented in lower field GF(2^2), the TI of the same is realized by replication of the sub-blocks, as shown in Fig. 2. The ANF expressions derived for the inverter are shown in Eq. (4). The inverter has an algebraic degree of "3" since the maximum number of variables in a minterm is three. Hence, the number of output shares computed for $d = 1$ using Eq. (3) for the inverter blocks is 8. Based on these values, the shared expressions of the inverter are shown in Eq. (5). The output shares of the multiplier, inverter, and S-box are further reduced to 2 so that they are compatible with the two-share top-level encryption/decryption data path.

To split the input data into two-shares, a PRNG based on the linear feedback shift register (LFSR) structure is employed. The LFSR's 128-bit output is used as one-share, and the second share is derived as the actual input XORed with this random number. This procedure is used for sharing both plaintext and secret key.

In order to satisfy the noncompleteness property, the S-box is pipelined to three stages, and for the uniformity property, a fresh randomness of about 12, 24, and 28 bits is introduced at each pipelining stage, as shown in a work [30]. The implemented AES S-box architecture is shown in Fig. 3. The two-share S-box architecture is embedded in the top-level RF-CF-S-box reuse-based architecture, which is modified to process the two shares of plaintext and key. The other linear blocks in

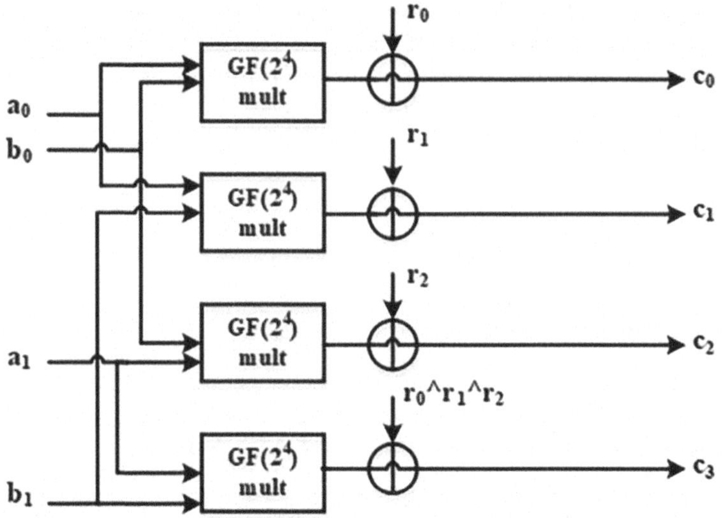

Fig. 2 Two-share TI of the GF(2^4) multiplier [30]

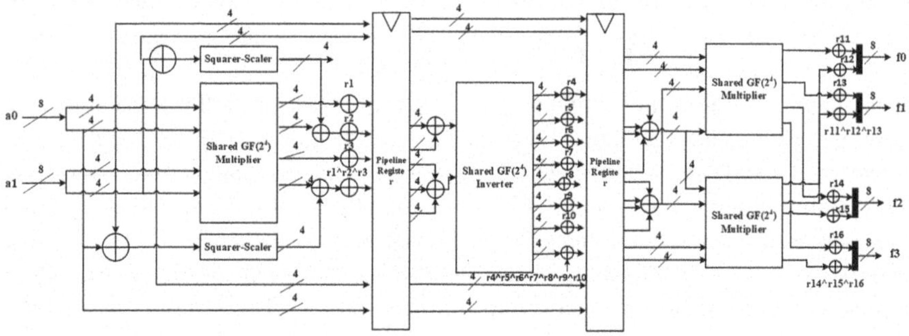

Fig. 3 Two-share 3-stage pipelined TI of the CFA-Based S-box [30]

the data path, namely XOR blocks and multipliers of the MixColumns block, are replicated twice to process the two shares. Thereby, a two-share AES architecture is developed, and the functional correctness is verified.

For the TI equations derived in this work, a[3:0] and b[3:0] represent the 4-bit inputs, a0[3:0] and a1[3:0], b0[3:0] and b1[3:0] represent the corresponding shares of the inputs. f[3:0] represents the 4-bit output, f0[3:0], f1[3:0], f2[3:0], f3[3:0], f4[3:0], f5[3:0], f6[3:0], and f7[3:0] represent the shares of the outputs.

$$f[0] = a[3] \oplus a[2] \oplus (a[3] \,\&\, a[0]) \oplus (a[3] \,\&\, a[2] \,\&\, a[1]);$$

$$f[1] = a[2] \oplus (a[3] * a[0]) \oplus (a[2] * a[1]) \oplus (a[3] * a[2] * a[1]) \oplus (a[3] * a[2] * a[0]);$$

$$f[2] = a[3] \oplus a[2] \oplus a[1] \oplus (a[2] * a[0]) \oplus (a[3] * a[2] * a[1]) \oplus (a[3] * a[1] * a[0]);$$

$$f[3] = a[2] \oplus a[1] \oplus a[0] \oplus (a[3] * a[1]) \oplus (a[3] * a[0]) \oplus (a[2] * a[1]) \oplus (a[3] * a[2] * a[1]) \oplus (a[3] * a[2] * a[0]) \oplus (a[2] * a[1] * a[0]) \oplus (a[3] * a[1] * a[0]); \tag{4}$$

$$f0[0] = (a0[3] * a0[2] * a0[1]) \oplus (a0[3] * a0[2] * a0[0]) \oplus (a0[3] * a0[1] * a0[0]) \oplus (a0[2] * a0[1] * a0[0]) \oplus (a0[3] * a0[1]) \oplus (a0[3] * a0[0]) \oplus (a0[2] * a0[1]) \oplus a0[2] \oplus a0[1];$$

$$f1[0] = (a0[3] * a0[2] * a1[1]) \oplus (a0[3] * a0[2] * a1[0]) \oplus (a0[3] * a1[1] * a1[0]) \oplus (a0[2] * a1[1] * a1[0]) \oplus (a0[3] * a1[1]) \oplus (a0[3] * a1[0]) \oplus (a0[2] * a1[1]) \oplus a0[1]; f2[0] = (a0[3] * a1[2] * a0[1]) \oplus (a0[3] * a1[2] * a1[0]) \oplus (a0[3] * a0[1] * a1[0]) \oplus (a1[2] * a0[1] * a1[0]) \oplus (a1[2] * a1[1]);$$

$$f3[0] = (a0[3] * a1[2] * a1[1]) \oplus (a0[3] * a1[2] * a0[0]) \oplus (a0[3] * a1[1] * a0[0]) \oplus (a1[2] * a1[1] * a0[0]) \oplus (a1[2] * a1[1]);$$

$$f4[0] = (a1[3] * a0[2] * a0[1]) \oplus (a1[3] * a0[2] * a1[0]) \oplus (a1[3] * a0[1] * a1[0]) \oplus (a0[2] * a0[1] * a1[0]) \oplus (a1[3] * a0[1]) \oplus (a1[3] * a1[0]);$$

$$f5[0] = (a1[3] * a0[2] * a1[1]) \oplus (a1[3] * a0[2] * a0[0]) \oplus (a1[3] * a1[1] * a0[0]) \oplus (a0[2] * a1[1] * a0[0]) \oplus (a1[3] * a1[1]) \oplus (a1[3] * a0[0]);$$

$$f6[0] = (a1[3] * a1[2] * a0[1]) \oplus (a1[3] * a1[2] * a0[0]) \oplus (a1[3] * a0[1] * a0[0]) \oplus (a1[2] * a0[1] * a0[0]);$$

$$f7[0] = (a1[3] * a1[2] * a1[1]) \oplus (a1[3] * a1[2] * a1[0]) \oplus (a1[3] * a1[1] * a1[0]) \oplus (a1[2] * a1[1] * a1[0]) \oplus a1[0];$$

$$f0[1] = (a0[3] * a0[2] * a0[1]) \oplus (a0[3] * a0[1] * a0[0]) \oplus (a0[2] * a0[0]) \oplus (a0[3] * a0[2]) \oplus a0[1];$$

$$f1[1] = (a0[3]*a0[2]*a1[1]) \oplus (a0[3]*a1[1]*a1[0]) \oplus (a0[2]*a1[0]) \oplus a1[1];$$

$$f2[1] = (a0[3]*a1[2]*a0[1]) \oplus (a0[3]*a0[1]*a1[0]) \oplus (a1[2]*a1[0]) \oplus a1[2];$$

$$f3[1] = (a0[3]*a1[2]*a1[1]) \oplus (a0[3]*a1[1]*a0[0]) \oplus (a1[2]*a0[0]);$$

$$f4[1] = (a1[3]*a0[2]*a0[1]) \oplus (a1[3]*a0[1]*a0[0]) \oplus a1[3];$$

$$f5[1] = (a1[3]*a0[2]*a1[1]) \oplus (a1[3]*a1[1]*a0[0]);$$

$$f6[1] = (a1[3]*a1[2]*a0[1]) \oplus (a1[3]*a0[1]*a1[0]);$$

$$f7[1] = (a1[3]*a1[2]*a1[1]) \oplus (a1[3]*a1[1]*a1[0]);$$

$$f0[2] = (a0[3]*a0[2]*a0[0]) \oplus (a0[3]*a0[2]*a0[1]) \oplus (a0[3]*a0[0]) \oplus (a0[2]*a0[1]) \oplus a0[2];$$

$$f1[2] = (a0[3]*a0[2]*a0[0]) \oplus (a0[3]*a0[2]*a0[1]) \oplus (a0[3]*a1[0]) \oplus (a0[2]*a1[1]);$$

$$f2[2] = (a0[3]*a1[2]*a0[0]) \oplus (a0[3]*a1[2]*a0[1]) \oplus (a1[2]*a0[1]);$$

$$f3[2] = (a0[3]*a1[2]*a1[0]) \oplus (a0[3]*a1[2]*a1[1]) \oplus (a1[2]*a1[1]);$$

$$f4[2] = (a1[3]*a0[2]*a0[0]) \oplus (a1[3]*a0[2]*a0[1]);$$

$$f5[2] = (a1[3]*a0[2]*a1[0]) \oplus (a1[3]*a0[2]*a1[1]);$$

$$f6[2] = (a1[3]*a1[2]*a0[0]) \oplus (a1[3]*a1[2]*a0[1]) \oplus (a1[3]*a0[0]);$$

$$f7[2] = (a1[3]*a1[2]*a1[0]) \oplus (a1[3]*a1[2]*a1[1]) \oplus (a1[3]*a1[0]) \oplus a1[2];$$

$$f0[3] = (a0[3]*a0[2]*a0[1]) \oplus (a0[3]*a0[0]);$$

$$f1[3] = (a0[3]*a0[2]*a1[1]) \oplus a0[3];$$

$$f2[3] = (a0[3]*a1[2]*a0[1]) \oplus a1[2];$$

$$f3[3] = (a0[3]*a1[2]*a1[1]) \oplus (a0[3]*a1[0]);$$

$$f4[3] = (a1[3]*a0[2]*a0[1]) \oplus (a1[3]*a0[0]);$$

$$f5[3] = (a1[3]*a0[2]*a1[1]) \oplus a0[3];$$

$$f6[3] = (a1[3]*a1[2]*a0[1]) \oplus a1[3];$$

$$f7[3] = (a1[3]*a1[2]*a1[1]) \oplus (a1[3]*a1[0]); \quad (5)$$

4.2 Design of AES Hardware Accelerator

The AES design has been implemented in such a way that all key variants, such as 128-, 192-, and 256-bit encryption and decryption, can be performed using the same design. The design implemented is a 32-bit iterative core architecture with an inbuilt key scheduling module and a register bank of 2 KB to store the round key values. The input–output data width is fixed at 128 bits. The block diagram of the AES hardware accelerator is shown in Fig. 4.

The top wrapper module (AES_INTERFACE_MODULE) is designed in such a way that a register bank is used to store the key and plain texts that are loaded from

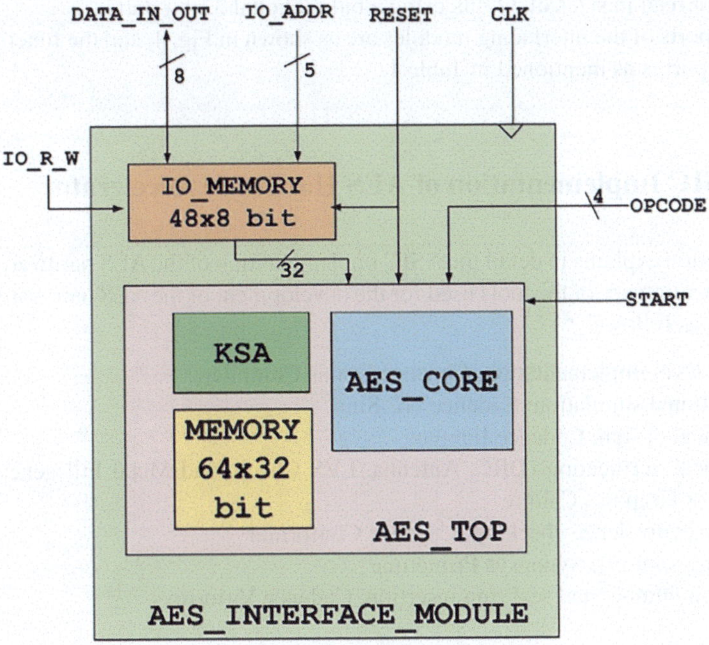

Fig. 4 Block diagram of the developed AES hardware accelerator

Table 1 I/O ports description

IO port	Description
IO_R_W	This input port is used to set the top module in read mode or write mode to read the output data or write the input data and key.
DATA_IN_ OUT [7:0]	This is a bidirectional port that is used to feed data, key in sizes of 8 bits, and retrieve back operated data through the same port.
IO_ADDR[4:0]	This input port is a 5-bit address feed to the register bank, denoting the address at which it has to read or load data.
RESET	This input signal is an active low signal, which resets the entire system.
CLK	This is the system clock fed to the entire design.
OPCODE[3:0]	This input signal is 4 bits wide and contains the MSB values for the mode of operation in which data has to be encrypted (128,192,256) using two bits, and the LSB bits denote the D_K signal, which denotes if data or a key is fed, and E_D which denotes if the operation to be performed is encryption or decryption.
START	This input signal is fed to the core, which indicates that the core can take the key or data that is stored in the register bank for operation, and this signal also starts the core operations like key scheduling algorithms and encryption/decryption.

the external environment, and the output of the algorithm, i.e., the encrypted data, is loaded back into the same register bank from which the output can be read. The main objective of the top wrapper module is to reduce the number of IO ports for the design. Since the core is a 32-bit iterative core architecture, data or key should be loaded or read in sizes of 32 bits using a bidirectional 32-bit port.

The ports of the interfacing modules are as shown in Fig. 4, and the functionality of each port is as mentioned in Table 1.

5 ASIC Implementation of AES Hardware Accelerator

This section explains in detail the ASIC implementation of the AES hardware accelerator. A summary of the tools used for the development of the AES hardware accelerator is as follows:

- Gate-level implementation: Cadence Genus Compiler
- Functional simulation: Cadence NC Sim
- Physical design: Cadence Innovus
- Physical verification (DRC, Antenna, LVS Check, and Metal Fill generation): Mentor Graphics Calibre
- Logic equivalence checking: Cadence Conformal
- Timing sign-off: Synopsys Primetime
- Silicon number and seal ring insertion: Cadence Virtuoso

5.1 AES Logic Synthesis

The logic synthesis of the design involves mapping the RTL code written in hardware description language to a standard cell-based netlist file based on the standard cells present in the technology library and constraints mentioned by the user. A technology library generally consists of gate-level technology-based definitions for different gates and macros. Logic synthesis primarily consists of three steps:

- Translation: RTL code is translated to technology-independent representation. The converted logic is available in Boolean equation form.
- Optimization: A Boolean equation is optimized using SoP or PoS optimization methods.
- Technology mapping: Technology-independent Boolean logic equations are mapped to technology-dependent library logic gates based on design constraints and the library of available technology gates. This produces optimized gate-level representation, which is generally represented in Verilog.
- The design is synthesized at three levels, and they are as follows:
- Generic synthesis: Converting RTL to gate-level netlist (syn_generic).
- Map synthesis: Mapping gates in the netlist to standard cells present in the technology library (syn_map).
- Opt synthesis: Optimizing the mapped netlist. (syn_opt).

These three synthesis commands can be performed at three levels of effort, such as low, medium, and high. This design was synthesized with medium effort. After optimizing the synthesis, the netlist file along with the synopsys delay format (SDC) file is dumped out, and these two files are used for post-synthesis simulation using Model Sim. The performance result of the synthesis process is shown in Table 2.

5.2 Physical Design

The outputs of logic synthesis are a technology-mapped netlist in Verilog format and a design constraints file in ".sdc" format. This Verilog netlist file contains the instantiation of the logic cells and their interconnection according to the source RTL synthesis and technology mapping based on the technology library files. Based on this netlist file, the actual layout design is developed using EDA tools like Synopsys IC Compiler, Cadence INNOVUS, etc. In this design, for the layout design of the AES, Cadence INNOVUS tool is used. The netlist is imported into the INNOVUS tool, and the layout design is done.

Table 2 Performance results of synthesis

Design	Area	Power	Gates count
AES–RTL synthesis result	506085.551 mm^2	69.5 mW	10,746

In the physical design flow, i.e., layout design, there are various steps involved. It is an iterative process in which the design steps have to be iterated by making appropriate changes in each iteration in order to meet the timing requirements and other design constraints like area, speed, and power. The following are the steps involved in the physical design:

- Floor planning
- Partitioning
- Power planning
- Placement
- Clock tree synthesis
- Routing
- Dummy metal fill
- Sign-off

After each step of physical design, timing analysis is performed to check whether the timing requirements are met and also to check for any design rule violations and optimize them at each step.

The final sign-off step involves static timing analysis (STA). Finally, when the design layout is complete, the place and route (PNR) tool generates a new netlist that has information about any modifications made to the original netlist, namely buffer addition, cell size changes, etc. It also creates a "definition" file, which has connectivity information between the logical cells, viz., wire length, width, locations, etc. This definition file is used to extract additional timing information due to wire-inbuilt RCs (resistances and capacitances) and store them into a separate file usually referred to as the SPEF (Standard Parasitic Extraction Format) file.

The sign-off step also involves physical verification and formal verification, in which the GDSII file is exported from the place and route, along with the final netlist, into the physical verification tool for verification. After the verification is successfully completed, the final GDSII file is sent to the foundry for fabrication.

The following are the components that have to be given to the PNR tool:

- Gate Level Netlist: It represents the technology mapped and synthesis netlist generated by the Logic Synthesis tool and contains the instantiation of standard cells based on the standard cell library provided during logic synthesis.
- Standard Cell Libraries: In digital design, the same standard cell library is used for both the synthesis and subsequent layout design. The netlist will have the instantiation of these cells. For digital layout, the layout and timing abstracts of the cells are used.
- Layout Model: An abstract model of the standard cell layout is used instead of the complete layout. Cadence EDI tools use layout exchange format (LEF) views. The LEF view is a cell view that has only the pins and blockages of cells defined.
- Timing Model: Tools also need a timing model in the form of a .lib file. This Liberty format file will have timing numbers for the various arcs in a cell,

generally in a look-up model. It is also important to note that .libs may also have cell power information.

- Technology Libraries: The rules pertaining to the process should also be given to the PnR tool. This includes metal widths, spacing, via definitions, etc.
- Timing Constraints: SDC files define the timing constraints of the design. It contains the clock definitions, false paths, any input and output delay constraints, etc.

These required constraints and libraries are read into the tool, and the design is initialized. In AES design, the netlist and SDC constraints from the logic synthesis are imported. The timing libraries and technology files (.lef) are available in the SCL pdk. Initially, the design process is set to 180 nm using the command setDesignMode process command. During the logic synthesis, high fanout nets like reset and scan enable are set as ideal networks because they are directly connected to all the flipflops. These nets are set as ideal in logic synthesis for analysis. So, such ideal network constraints have to be removed before PNR flow.

5.2.1 Floor Planning

This is the first step of the layout design, in which the position of the IOs, Macros, and design modules is determined. This is the major step in the layout design. The floor plan of the design determines the chip quality. At this step, the size of the chip block, the allocation of power routing resources, the placement of the hard macros, and the space reservation for standard cells are completed. In the case of the AES design, a number of iterations were performed to find the optimum core size. The IO pads are fixed based on the analysis and routing congestion in the trial run.

- Core boundary: It refers to the area where the standard cells and their IP blocks are placed. Power routing spaces fall outside the core boundary. For a full chip, IO buffers and IO pads are placed outside the core boundary. In the PnR tool, the floor planning can be controlled by various parameters:
- Aspect ratio: It is defined as the ratio of height to width, and it is useful to determine the shape of the floor plan as square or rectangular. An aspect ratio of 1 results in a square-shaped floor plan.
- Core utilization = (standard cell area + macro cells area)/total core area.
- A core utilization of 0.95 implies that 90% of the area is available for the placement of cells, whereas 0.05% is left free for routing.
- Boundary: Boundary is another control parameter, specifying which makes the h aspect ratio and core utilization to be irrelevant.
- In the AES layout design, the aspect ratio was set to 1, and the core utilization was set to 80 percent so the remaining space could be used for routing. Also, in AES design, there are no hard macros to be placed, so this makes floor planning much easier. Hence, the core area is completely used for the standard cell placement. Each standard cell row is 5.6 μm in height, and the spacing between each row is 1.12 μm. The rows are not flipped to each other and are not abutted. Totally, there are 150 standard cell rows in the core area for standard cell

placement. And an offset of 20 μm from each side of the core boundary is left for power ring creation. Hence, the IO pads are placed at 20 μm offset from the core boundary. The floor plan of the completed design is shown in Fig. 5.

The area obtained in Fig. 5 after the floor plan is 995.651 mm^2, and the number of instances has increased due to the addition of buffers and fillers to 14,522.

2. IO Planning: IO planning is the placement of IO pad or pins. Once the IO positions are fixed by the designers, the IO pads are placed on all four sides of the core, based on the requirements along with the design size and aspect ratio of the design. In this work, the IO placement is done by using an IO file in which the position of the IO pads, the orientation of the pads, and the spacing between the pads are specified. This file is read into the EDI tool before initializing the design while importing it.

5.2.2 Power Planning and Power Routing

The power planning step involves the creation of power rings and power stripes. The power rings and stripes are created based on the power requirements of the chip. There are two power domains used in the AES design, namely core power and IO power. The power ring (VDD VSS) around the core boundary includes the core power rings and the IO power rings (VDDO VSSO), which are formed automatically on the placement of IO pads and IO filler pads. The number of IO power pads and core power pads also depends on power requirements. They can be manually calculated by using the power of the chip. The total power of the chip is obtained

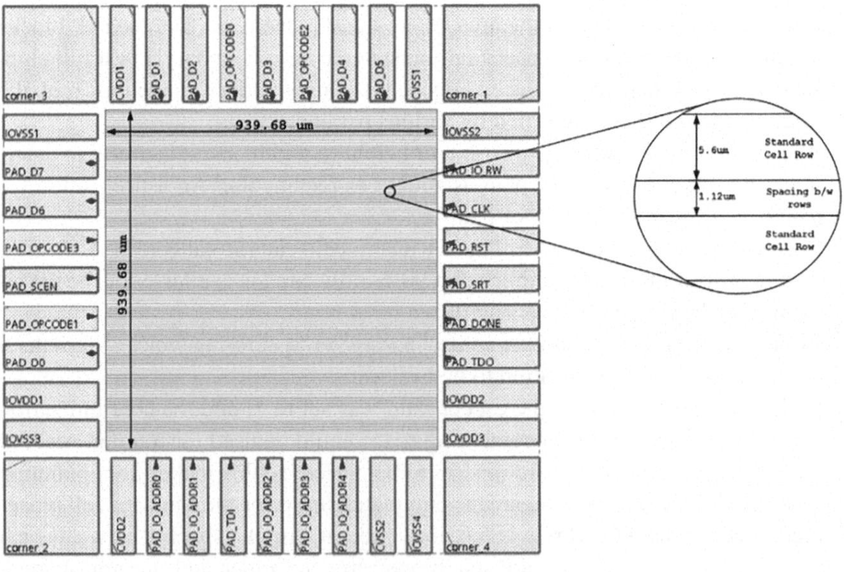

Fig. 5 Floorplan of AES hardware accelerator

from the logic synthesis; the accurate power can be obtained by using an activity file that can be created in the RTL simulation in any simulation tool.

For AES power planning,

- Number of core power pads: 2 sets (VDD VSS)
- Number of IO power pads: 2 sets (VDDO and VSSO)
- Core power (report of a rough run of physical synthesis): 36.44 mW
- Core voltage: 1.8 V

Total dynamic core current (mA):

= total dynamic core power/core voltage.
= 36.44/1.8
= 20.24 mA

Core PG ring width (both VDD and VSS) = (total dynamic core current) / (no. of sides * maximum current density of the metal layer used (Jmax) for the PG ring).

= 20.24 mA / (4 * 1 mA/μm) = 5.06 ~ (5 μm)

No. of vertical stripes: 3 sets at an offset of ~250 μm.

The vertical power stripes created should be larger in width when compared to the horizontal power stripes of standard cells. After the creation of power rings and stripes, these global nets are connected to the VDD, VSS, VDDO, and VSSO pins. After the global nets are connected, power routing, namely special routing, is performed. After the power net routing, all power nets of standard cell rows get tapped to the power rings, and the power pin of the power pads gets routed to the core rings.

Now, the IO filler cells are added to fill the gap between the IO pads; the IO fillers are added in such a way that there is no overlap between the cells. The IO filler cells that are used are pfeed30000, pfeed10000, pfeed02000, pfeed01000, pfeed00120, and pfeed00010. The placement of the cell automatically forms the IO PG rings. Fig. 6 shows the AES layout after power planning.

5.2.3 Placement

After the power routing is completed, the next step in the physical design flow is the placement of the cell. In this step, the cells are placed in the standard cell rows that were created during the floor plan. In the AES design, before placement, a design exchange format (DEF) file is imported, which defines the scan chain connections of the design. The DEF file is a universal format that provides compatibility of design between different tools.

After placement, the timing is analyzed and also checked for design rule violations (DRV) such as max_capacitance, max_transition, and max_fanout. Then, the design is optimized to fix the DRV violations, such as high fanout nets. The setup timing violations of the AES design are fixed during the placement stage, and the hold timing had a negative slack of −0.34; this hold violation can be fixed after the clock tree synthesis. The placed design of AES is as shown in Fig. 7.

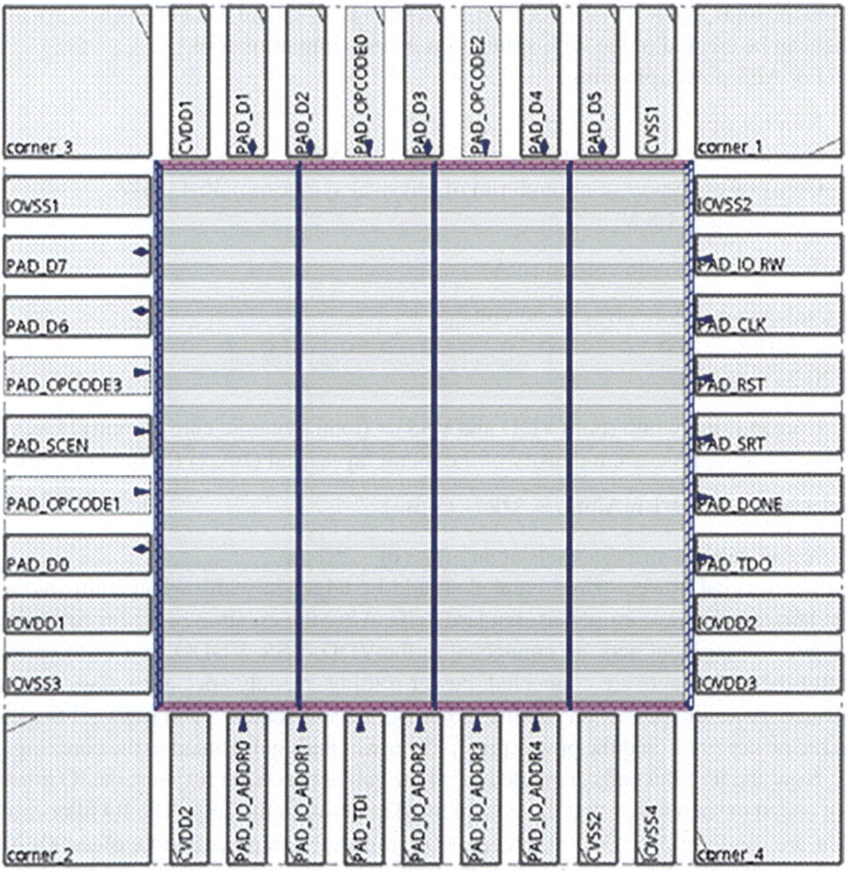

Fig. 6 AES design after power planning

5.2.4 Clock Tree Synthesis

A clock signal is used to synchronize the data transfer between functional elements. The clock sources can be internal, namely PLLs or oscillators within the chip, or external, connected through an IO namely a clock pin. It is important to do clock balancing to meet the required design constraints, and this can be achieved by clock tree synthesis (CTS) after the placement step.

Figure 8 (a) shows how a clock net is present before the CTS, and Fig. 8 (b) shows the same after the CTS.

The main concerns to be addressed in CTS are.

- Skew: It is important to reduce clock skew. The skew is defined as the maximum difference among the delays from the clock source to the clock sink.
- Clock Source: Specified using the "create_generated_clock" command.

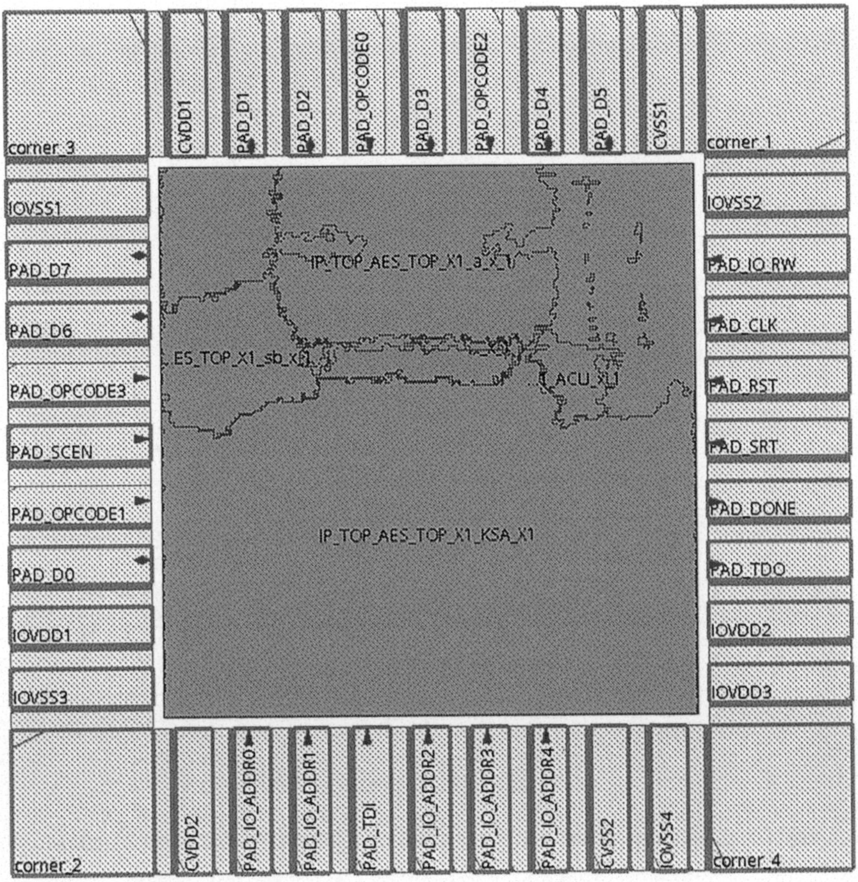

Fig. 7 AES design after placement

- Clock Sinks: Sinks are the nodes that receive the clock. Default sinks are the clock pins of flipflops.
- CTS should aim to minimize the skew in the design so that every clock sink receives the clock at the same time. After the completion of CTS, timing needs to be analyzed to fix any setup or hold violations.

5.2.5 Routing

Routing is the subsequent process after CTS to perform the required interconnections. This process involves making the required electrical connection through metal and vias on the standard cell, its macro pins, the pins on the block boundary, or pads at the chip boundary. It is important to ensure the following:

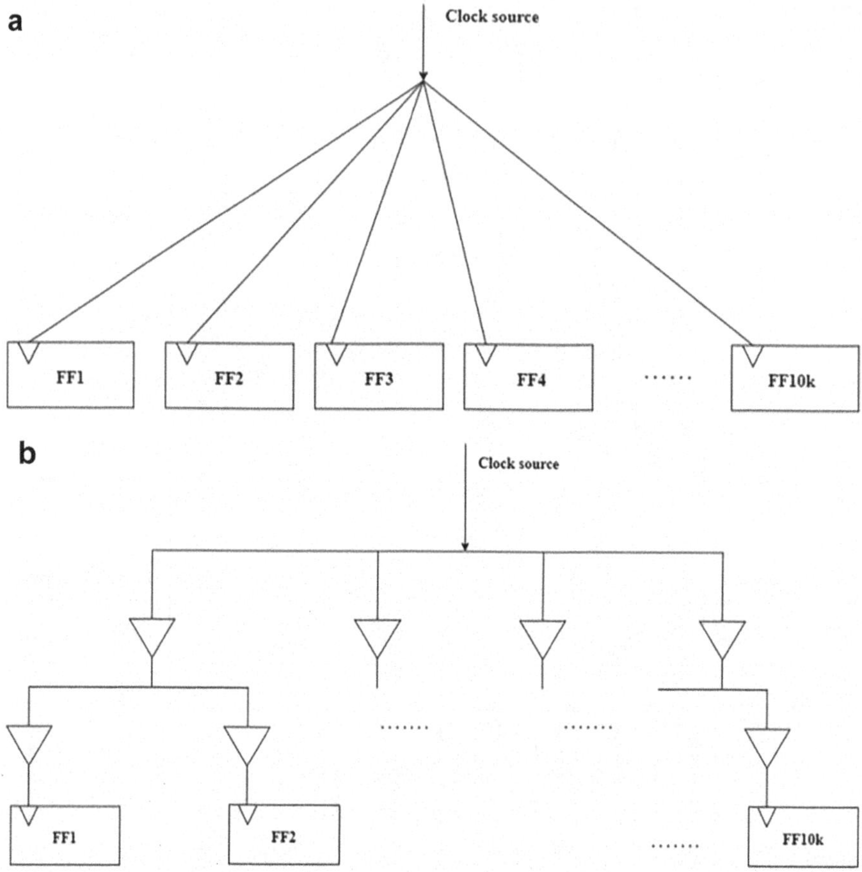

Fig. 8 (**a**) Clock net before CTS. (**b**) Clock net after CTS.

- The tool should complete all the connections specified by the netlist without any LVS errors.
- Routing should adhere to the design rules.
- All the timing constraints need to be met.

The width and spacing for each layer and via are mentioned in the technology file (LEF). Design Rule Checks (DRCs) ensure that the required parameters, like spacing, width, via size, etc., follow the manufacturing requirements to minimize yield.

The routing process is done at various stages of the PnR flow, namely global routing, track assignment, and detailed routing. In global routing, the design is partitioned into different routing regions. Routing congestion is also determined by the following factors:

- The number of nets that should pass through the region.
- The number of routing tracks that are available in the region.

In detailed routing, global routing results are used to lay the actual wires interconnecting the nodes. It is important to analyze congestion and fix it before doing detailed routing.

5.2.6 Sign-Off

After creating a layout, post-routing and optimizing the layout design in the PNR tool, the GDSII file of the design is dumped out. The GDSII file is further imported into a layout editor tool. In the layout editor tool, design rule checks (DRC) and layout vs. schematic (LVS) checks are run. After clean DRC and LVS checks, the parasitic extraction is run, which produces a pex.netlist file or a .spef file. This file specifies the parasitic (RC) values of each entity in the design. The SPEF file or PEX file is used to calculate the delays of each interconnect in the design in the timing sign-off step.

The de facto industry standard that is adopted in industry for data exchange of IC layout is GDSII. GDSII represents the IC layout in planar geometric shapes, text labels, and other information in binary format. The final output format of the PnR flow is the GDSII files, which are further given to silicon foundries for IC fabrication. It consists of four parameters, namely.

- Layer object name
- Layer object type
- Layer number
- Data type

In the AES design, the GDSII file was streamed out from the Innovus PNR tool. This GDSII file was imported as a cell in a layout editor tool called Cadence Virtuoso. In Virtuoso, after importing the GDSII file, DRC and LVS checks were performed. After clean DRC and LVS, the parasitic extraction step is done, which dumps out a. SPEF file. This SPEF file is the input for the timing sign-off tool, which calculates the net delays using parasitic values. The final completed layout is shown in Fig. 9.

A timing sign-off is performed to analyze whether the timing requirements are met for the final layout of the design. For timing sign-off, the inputs are the SPEF file, the post layout netlist file (.v), the constraints file (SDC) given by the user, which mentions the target clock skew, the input and output delays, and the technology library. In timing sign-off, tools check the hold and setup timing, clock skew, and calculate annotated delays for each net using the SPEF file. Finally, an SDF file can be written using this tool that can be used for post-layout simulation, from which the net and cell delays can be annotated. The netlist file dumped by the PNR tool post-routing can be used to simulate the design in any simulation tool, along with the delays annotated by the SDF file, which is produced by the timing sign-off tool. Using this, the functionality of the layout design is verified in a simulated environment. For AES design, timing sign-off was performed using the Synopsys

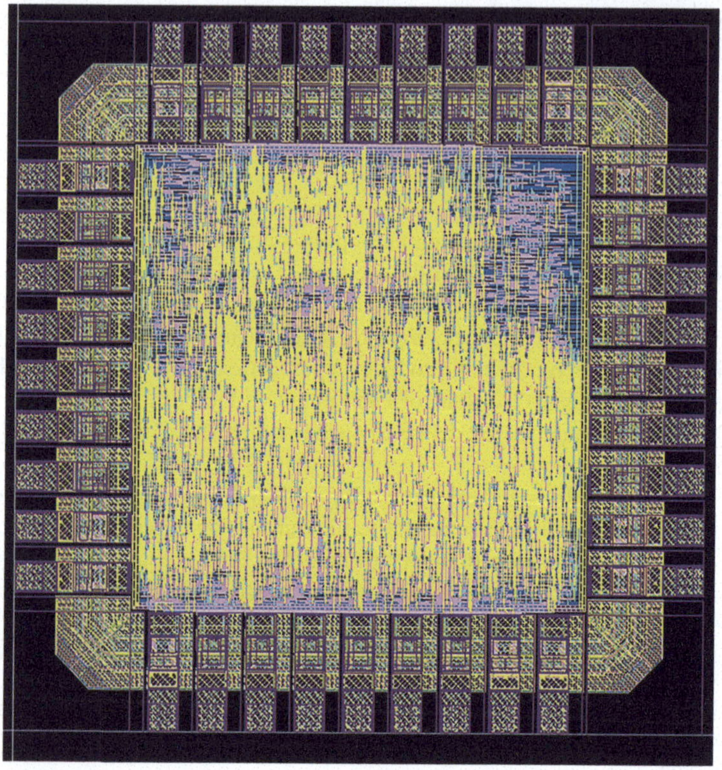

Fig. 9 Layout of the AES design

Prime Time tool, and the post-layout simulation along with the SDF file was performed in the ModelSim simulator tool.

After the timing is met and the post-layout simulation is successful, the layout design is not modified. In the layout editor tool, i.e., in Virtuoso, seal rings, silicon number, and dummy fills are added to the layout design. The seal ring is an outer ring that covers the entire layout. This seal ring is placed in order to mention that the wafer has to be cut to get the die of our design after fabrication. The silicon number is a convention to identify the die in the wafer, and it is unique for every wafer fabricated. Dummy fills are metal fills over the design inside the sealring and outside the core's layout that fills the unused gaps with active metal fills. The addition of dummy metal fill increases the reliability of the fabrication process. Finally, GDSII is exported along with dummy metal fills, a silicon number, and seal ring. This GDSII can be used to proceed with the fabrication process of the design. The layout with the seal ring is shown in Fig. 10.

The final completed layout and package information are sent to SCL Foundry for fabrication, and the taped-out AES IC is shown in Fig. 11.

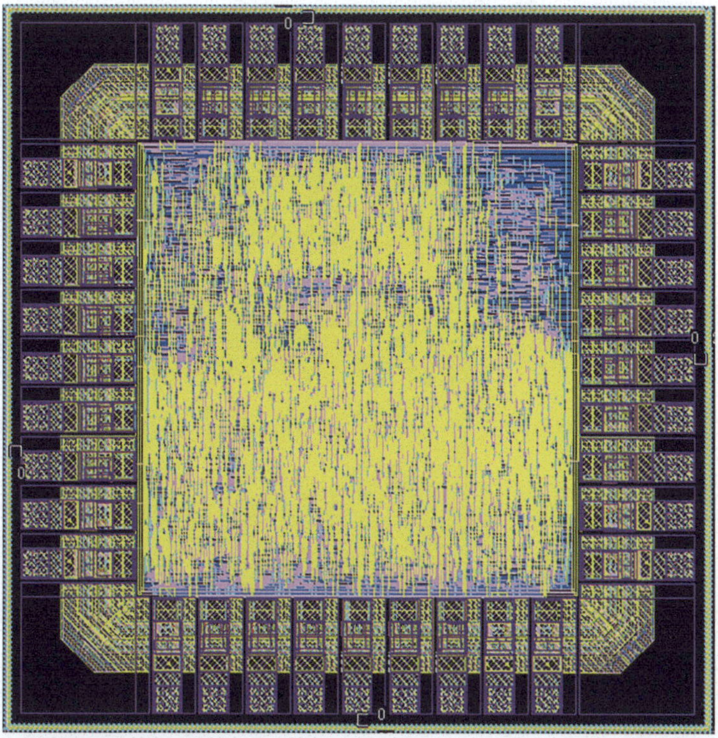

Fig. 10 Final layout of AES design with the seal ring

6 Performance Analyses

This section shows the comparison of the ASIC implementation of the AES algorithm on a 180 nm technology node with existing works in the literature. The works described in [35–37] are unprotected AES implementations, and the work in [38] is a masked implementation of AES. As seen in Table 3, the developed AES hardware accelerator has achieved the lowest cost and the lowest power, with a reasonable throughput of 5 Mbps suitable for security-critical applications.

7 Conclusion

The ASIC implementation of power analysis attack-tolerant AES has been designed and implemented. A two-share-based TI is developed and implemented such that the developed AES hardware accelerator has the lowest cost and a reasonable throughput when compared with existing works in the literature. The ASIC implementation process is explained in detail, along with the two-share TI equation

Fig. 11 Fabricated IC of AES hardware accelerator

Table 3 Comparison of AES implementations on ASIC 180 nm technology

Reference	Technology	Area (gate count)	Clock frequency	Power	Throughput
LUT based [35]	Synopsys 180 nm	3.96 mm^2	100 MHz	–	1.82 Gbps
Fully integrated [36]	TSMC 180 nm	28,626	465 MHz	0.314 W	1.16 Gbps
Flexible AES [37]	CMOS 180 nm	19,500	100 MHz	–	0.85 Gbps
Masked AES [38]	Samsung 180 nm	20,506	5 MHz	2.0 mA	4 Mbps
This work	SCL 180 nm	14,522	50 MHz	36.44 mW	5 Mbps

derivation. The developed AES hardware accelerator has achieved the lowest cost and the lowest power, with a reasonable throughput of 5 Mbps suitable for security-critical applications. The taped-out AES ASIC can be used in security-critical areas like IoT edge devices, where a dedicated ASIC will speed up the processing of the entire system. The future work will aim to reduce the amount of randomness used for two-share-based TI since the technique requires about 64 bits of randomness per S-box, apart from the randomness required for the input and key masking. The future work will also develop test beds suitable for testing the working of the crypto accelerator.

References

1. National Institute of Standards and Technology (NIST), advanced encryption standard, Federal Information Processing Standards Publication 197 (2001), Available at. http://csrc.nist.gov/publications/ps/ps197/ps-197.pdf.
2. Chellam, M. B., & Natarajan, R. (2017). AES hardware accelerator on FPGA with improved throughput and resource efficiency. *Arabian Journal for Science and Engineering, 43*(1), 18.
3. Cao, Q., & Li, S. (2009). *A high-throughput cost-effective ASIC implementation of the AES algorithm* (pp. 805–808). IEEE 8th international conference on ASIC.
4. Sharma, V. K., Kumar, S., & Mahapatra, K. K. (2016). Iterative and fully pipelined high throughput efficient architectures of AES in FPGA and ASIC. *Journal of Circuits, Systems and Computers, 25*, 1650049.
5. Kocher, P., Jaffe, J., & Jun, B. (1999). Differential power analysis. In *Annual international cryptology conference* (pp. 388–397). Springer.
6. Beyne, T., Dhooghe, S., Moradi, A., & Shahmirzadi, A. R. (2022). Cryptanalysis of efficient masked ciphers: Applications to low latency. *IACR Transactions on Cryptographic Hardware and Embedded Systems, 2022*(1), 679–721.
7. Masoumi, M. (2019, October). A highly efficient and secure hardware implementation of the advanced encryption standard. *Journal of Information Security and Applications, 48*, 102371.
8. Wang, H. (2024). Amplitude-modulated EM side-channel attack on provably secure masked AES. *Journal of Cryptographic Engineering, 14*, 537.
9. Ghosal, A. K., Sardar, A., & Chowdhury, D. R. (2024). Differential fault analysis attacktolerant hardware implementation of AES. *The Journal of Supercomputing, 80*(4), 4648–4681.
10. Akkar, M. L., Bevan, R., Dischamp, P., et al. (2000). Power analysis, what is now possible. In proceedings of the international conference on theory and application of cryptology and information security(pp. 489 502).
11. Singha, T. B., Palathinkal, R. P., & Ahamed, S. R. (2023). Securing AES designs against power analysis attacks: A survey. *IEEE Internet of Things Journal, 10*(16), 14332–14356.
12. Yu, W., & K se, S. (2017). A lightweight masked AES implementation for securing IoT against CPA attacks. *IEEE Transactions on Circuits and Systems I: Regular Papers, 64*(11), 2934–2944.
13. Zhao, X. (2024). Secure and efficient masking of lightweight ciphers in software and hardware. *The Computer Journal, 67*(2), 581–603.
14. Zhang, T., Tehranipoor, M., & Farahmandi, F. (2024). Trust guard: Standalone FPGA-based security monitoring through power side-channel. *IEEE Transactions on Very Large Scale Integration (VLSI) Systems, 32*(2), 319.
15. Yang, M., Ahmed, T., Inagaki, S., Sakiyama, K., Li, Y., & Hara-Azumi, Y. (2024). Hardware/software cooperative design against power side-channel attacks on IoT devices. *IEEE Internet of Things Journal, 11*(9), 16758–16768.
16. Song, J., Lee, K., & Park, J. (2022). Low area and low power threshold implementation design technique for AES S-box. *IEEE Transactions on Circuits and Systems II: Express Briefs, 70*(3), 1169–1173.
17. Yao, F., Chen, H., Wei, Y., Pasalic, E., Zhou, F., & Fan, L. (2024). Optimizing AES threshold implementation under the glitch-extended probing model. *IEEE Transactions on Computer Aided Design of Integrated Circuits and Systems, 43*, 1984.
18. Shahmirzadi, A. R., & Moradi, A. (2021). Re-consolidating first-order masking schemes. *IACR Transactions on Cryptographic Hardware and Embedded Systems, 2025*(1), 305–342.
19. Gigerl, B., Klug, F., Mangard, S., Mendel, F., & Primas, R. (2023). Smooth passage with the guards: Second-order hardware masking of the AES with low randomness and low latency. *IACR Transactions on Cryptographic Hardware and Embedded Systems (TCHES), 2024*(309), 335.
20. Dhooghe, S., & Ovchinnikov, A. (2023). Threshold implementations with non-uniform inputs. In *International conference on selected areas in cryptography* (pp. 97–123). Springer.

21. Shao, T., Bohua Wei, Y. O., Wei, Y., & Wu, X. (2023). New second-order threshold implementation of Sm4 block cipher. *Journal of Electronic Testing, 39*(4), 435–445.
22. Ng, J.-S., Chen, J., Kyaw, N. A., Chong, K.-S., & Gwee, B.-H. (2024). Securing against sidechannel attacks with wide-range in situ random voltage dithering on async-logic AES engine. IEEE transactions on very large scale integration (VLSI) SYSTEMS, *10*(32), 1959–1963.
23. Belleville, N., & Masure, L. (2024, April). Combining loop shuffling and code polymorphism for enhanced AES side-channel security. In COSADE 2024 15th international workshop on constructive side-channel analysis and secure design(pp. 260 280).
24. Chari, S., Jutla, C. S., Rao, J. R., et al. (1999). *Towards sound approaches to counteract power-analysis attacks. In proceedings of the annual international cryptology conference* (pp. 398–412). Springer.
25. Messerges, T. S. (2000). Securing the AES finalists against power analysis attacks. In proceedings of the 7th international workshop on fast software encryption(pp. 150 164).
26. Nikova, S., Rechberger, C., & Rijmen, V. (2006). Threshold implementations against sidechannel attacks and glitches. In international conference on information and communications security(pp. 529 545).
27. Moradi, A., Poschmann, A., Ling, S., Paar, C., & Wang, H. (2011). Pushing the limits: A very compact and a threshold implementation of AES. In annual international conference on the theory and applications of cryptographic techniques (pp. 69 88).
28. Bilgin, B., Gierlichs, B., Nikova, S., Nikov, V., & Rijmen, V. (2015). Trade-offs for threshold implementations illustrated on AES. *IEEE Transactions on Computer-Aided Design of Integrated Circuits and Systems, 34*(7), 1188–1200.
29. Cnudde, D., Thomas, O. R., Bilgin, B., & Nikova, S. (2016). Ventzislav Nikov, and Vincent Rijmen: Masking AES with d+ 1 shares in hardware. In *International conference on cryptographic hardware and embedded systems* (pp. 194–212). Springer.
30. Ueno, R., Homma, N., & Aoki, T. (2017). Toward more efficient DPA-resistant AES hardware architecture based on threshold implementation. In *International workshop on constructive side-channel analysis and secure design* (pp. 50–64). Springer.
31. Daemen, J. (2017). Changing of the guards: A simple and efficient method for achieving uniformity in threshold sharing. In *International conference on cryptographic hardware and embedded systems* (pp. 137–153). Springer.
32. Wegener, F., & Moradi, A. (2018). A first-order SCA resistant AES without fresh randomness. In *International workshop on constructive side-channel analysis and secure design* (pp. 245–262). Springer.
33. Papagiannopoulos, K. (2018). Low randomness masking and shuffling: An evaluation using mutual information. In IACR transactions on cryptographic hardware and embedded systems(pp. 524546).
34. Sugawara, T. (2018). 3-share threshold implementation of AES s-box without fresh randomness. *IACR Transactions on Cryptographic Hardware and Embedded Systems, 2019*(1), 123–145.
35. Kuo, H., & Verbauwhede, I. (2001). Architectural optimization for a 1.82 Gbits/sec VLSI implementation of the AES Rijndael algorithm. In *Lecture notes in computer science, cryptographic hardware and embedded systems CHES* (Vol. 2162, pp. 51–64). Springer-Verlag.
36. Kim, N. S., Mudge, T., & Brown, R. (2003). Brown: A 2.3 Gb/s fully integrated and synthesizable AES Rijndael core (pp. 193 196). IEEE custom integrated circuits conference.
37. Li, H. (2006). Efficient and flexible architecture for AES. *IEE Proceedings-G: Circuits, Devi ces and Systems, 153*, 533–538.
38. Trichina, E., Korkishko, T., & Lee, K. H. (2004). Small size, low power, side channel-immune AES coprocessor: Design and synthesis results. In *International conference on advanced encryption standard* (pp. 113–127). Springer.

Design and Analysis of a Honeypot Pi for Intrusion Detection System

D. Saravana Kumar, Alicia Biju, Raja Muthalagu, Pranav M. Pawar, and Abdul Rajak

1 Introduction

A honeypot is a network-attached device that serves as a decoy to attract attackers and to track and investigate hacking attempts that obtain illegal access to digital systems. Honeypot solutions use hardened operating systems, which provide additional security features to reduce vulnerability to threats. Honeypot acts as a sacrificial computer and can be considered as one of the oldest security measures in the cybersecurity discipline [1–5]. The cost of a honeypot is expensive; however, it is reduced immensely when it is deployed onto a Raspberry Pi.

Honeypots positioned in a network's Demilitarized Zone (DMZ), as shown in Fig. 1, eliminate them from the production network while still keeping it linked. This helps reduce the chances of the system being compromised by a hacker. Attacking the internal networks is prevented when it is placed outside the external firewall; however, it cannot trap and identify attackers that are already inside [6–8]. When present in the internal network, it can detect internal attacks and misconfigured firewalls. A honeypot's placement depends on its intricacy, type of traffic, and closeness to organizational assets [9].

D. Saravana Kumar (✉) · A. Biju · R. Muthalagu · P. M. Pawar · A. Rajak
Birla Institute of Technology and Science Pilani, Dubai Campus,
Dubai, United Arab Emirates
e-mail: saravana.kumar@dubai.bits-pilani.ac.in; f20170025d@dubai.bits-pilani.ac.in; raja.m@dubai.bits-pilani.ac.in; pranav@dubai.bits-pilani.ac.in; abdulrazak@dubai.bits-pilani.ac.in

© The Author(s), under exclusive license to Springer Nature Switzerland AG 2025

A. Mishra et al. (eds.), *Hardware Security: Challenges and Solutions*,
https://doi.org/10.1007/978-3-031-81213-2_6

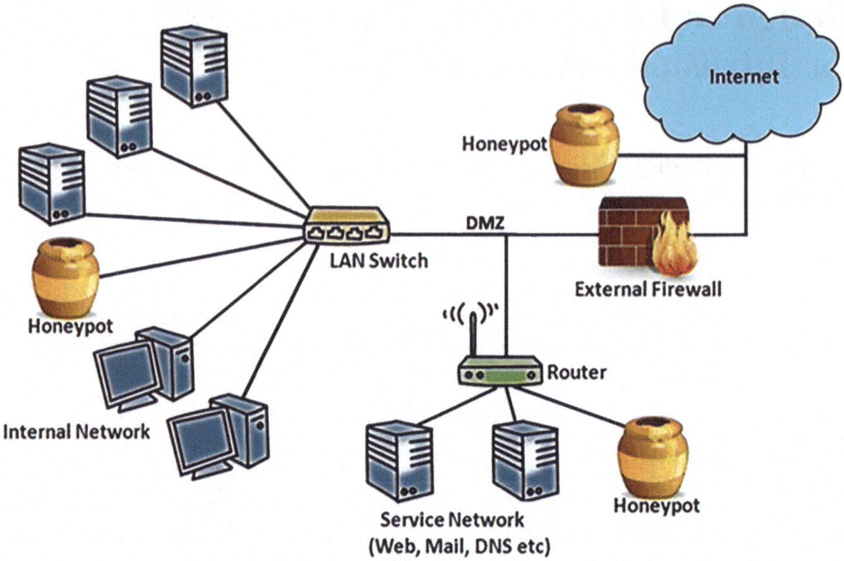

Fig. 1 Working model of a honeypot [10]

2 Literature Review

The paper titled [11] "Securing a Small Network by Using Raspberry Pi Honeypot," written by the authors Monica Catherine and Ashok Kumawat, discusses about how a home network or Internet of Things devices can be safeguarded by using a honeypot which will be deployed on a Raspberry Pi that will run an Intrusion Detection Program. The Intrusion Detection System uses a sensor called SNORT which analyzes raw live traffic and monitors these streams. Tshark is set up as the packet analyzer for sniffing and viewing the network packets for the user in a simple format. Cowrie and Dionaea are designed to be the honeypots. The honeypot would be implemented on the network to act as a decoy designed for attracting attackers and collecting details about both the intruder and the attack. The Cowrie medium-level interface honeypot has been used to monitor the full-shell interface during threats along with SSH and brute force attempts. This honeypot helps any intruder to start a communication and then sign further into the network, conning themselves into assuming they are accessing the legal SSH session of the target device. Any suspicious traffic, if discovered by the honeypot, will then help alert the user by sending a mail or a notification. Dionaea, the low-level interaction honeypot, is used to capture malware-leveraging weaknesses revealed by networked services. These captured logs are then stored into a server locally which is running on a Virtual Machine. They conclude that using the Raspberry Pi as a network monitoring system for protecting against security threats in a small network is less expensive than deploying the monitoring system on a single system [6].

The author Alexander Vetterl in his paper [12] "Honeypots in the Age of Universal Attacks and the Internet of Things" creates a honware which is a honeypot framework for speedy execution and utilization of high-level interaction honeypots. It instinctively manages a standard firmware picture and can mimic a variety of various devices without any entry into the manufacturer's hardware. Four case studies were taken by the author to demonstrate that honware is the first adaptable and standard framework to deploy honeypots inexpensively and successfully and can thus attain the precise data of attacks along with malware samples. The honware utilizes device firmware illustrations provided by CPE and IoT manufacturers, utilizing Quick Emulator (QEMU) to operate code for different CPU constructions on a single host machine and a unique prebuilt Linux kernel to imitate the device behavior inside a virtual environment. He concludes by saying that honware is a key step advancing in rebalancing the economics of assailants and guardians by cutting the assailants' skill to exploit weaknesses, predominantly "zero day" liabilities, for significant periods while the protectors are incapable to capture the specifics of the attack and thus begin the procedure of mitigation [11].

"Intrusion Detection Using Raspberry Pi Honeypot (Snort) For Network Security" is a research paper written by Bharathi E., Keerthana M., and Ramsundar G., which talks about how the intrusion detection technology with the help of a honeypot alongside Raspberry Pi can make the network security robust as it is simple to execute and cost-effective. The security system will be using an Intrusion Detection System combination with a Raspberry Pi honeypot where SNORT will be used as the tool for intrusion detection. SNORT is a packet sniffer that summarizes and checks network traffic in real time, analyzing every packet carefully to identify an interruption. It is a solution to reduce breakdowns in the recognition process and compilation of the vital data focused on the honeypot. The Raspberry Pi honeypot is an implementation of the client–server construction. It has a predominant key server cooperating with numerous clients in the network. It will be implemented with Dionaea or Glastopf or Kippo which will accumulate all the information and transmit it to the server. The Raspberry Pi–Honeypot as a trap in the network signifies a straightforward and an inexpensive solution for improving network security as this system combines the security tools to reduce the disadvantages and expand the security resources in the procedure of safeguarding the network [13].

The researchers Meng Wang, Javier Santillany, and Fernando Kuipers discuss about an IoT honeypot called ThingPot in the paper "ThingPot: an interactive Internet-of-Things honeypot". The ThingPot concentrates not only on the IoT application protocols but also on the overall IoT platform. ThingPot can be considered as a medium-interaction honeypot or a hybrid communication IoT platform honeypot, whose platform consists of XMPP/MQTT as high-interaction honeypot modules, whereas low-interaction honeypot device imitation is performed through a REST API. ThingPot replicates the frontend, backend, IoT devices, and present XMPP/MQTT services (servers, clients, and libraries). REST is used to construct a backend API. XMPP has been selected as the IoT protocol for immediate interaction. The frontend is recognized through a simple HTTP web service. The Philips Hue (Phue) which is a smart lighting system IoT product has been mimicked by the

ThingPot. The authors have supplied a Proof-of-Concept (PoC) application of ThingPot, which is made accessible as an open-source code as well, concentrating on the IoT platform by using Philips Hue, and the PoC had been utilized and seized real data for 1.5 months [14].

The author, Asonganyi, for this chapter discusses about the implementation of HoneySMB which is a honeypot for the SMB (Server Message Block) protocol, the HoneyWEB with SQL-injection weakness, and the HoneyDB which is a honeypot for the MySQL database. The author has also explained the design, implementation, and deployment of these honeypots and has also given an evaluation of the composed threat intelligence; the chapter also comprises further work on a new HoneyClient which is a client honeypot and a way to break into the Android Sandboxing environment. Honeypots give us different kinds of brute-forcing scripts, different kinds of malicious and exploit kits which can be useful in making our own system more robust against these kinds of malicious attacks. He concludes by saying that honeypots are useful in terms of defining the boundaries of firewalls, that is, how should one tighten their firewall rules to ensure secure communications through the Internet. ClientPot provides protection from malicious webservers, and it is used to detect drive-by-download malwares and helps make the system safe from these activities [15].

The article by the researchers [16], Neil Eliot, David Kendall, et al. have explained how a flexible laboratory is enough to support the training of cybersecurity through the secure incorporation of honeypot appliances, and it can give an analysis of the honeypot architectures. The small-scale High Interaction Low Volume (HILV) honeypots and the Low Interaction High Volume (LIHV) honeypots have been deployed for this research paper. The HILV honeypots are miniature-scale and low-cost, and the apparatus is eternally designed for educational purposes. Each HILV grants convenient cybersecurity components to be distributed more successfully. The use of Raspberry Pi boards in the HILV honeypots has permitted varied subjects to be lectured more effortlessly due to their use of detachable media to store the operating system. The incorporation of the honeypots into a workshop environment that encourages a broad range of other technology-centered components offers a cost-valued answer to the distribution of stimulating, practical cybersecurity teaching [14].

The study recommends an examination standard for gathering network intrusion proof to distinguish and examine the provided loophole within the network security. To execute this model, a research which includes the use of a rule-based methodology and Raspberry Pi model as a honeypot system to aid in data compilation was performed. The rule-based report is implemented for detecting malicious behaviors, and these events are recorded into a database. From the gathered data, individuals will monitor the trend and activity rates for the infringement activities by examining the statistical report, and so a shield methodology can be made before the assault is instigated. Execution of the DiniB (Detection and Investigation of Network Intrusion Behavior) method supports in analyzing the network intrusion behavior inside a host-based network. The proposed model intends to resolve the stated challenge statement and lessens the effect from the network intrusion behavior spreading throughout the network. In response to this, a forecast was produced for guessing the existence and amount of the intrusion incidents in advance, centered on the activity rates and designs detected from the information accumulated by using the Raspberry Pi model as a honeypot system [17].

Researchers Rodney Anthony Raj and Chayapathi A. R. discuss about how honeypots can be used to find out who the attackers are and the ways they use to get into a small network. They have proposed a system that can hold all records and communications of the network that can be examined to have the details of the intruder or utilize it as an investigative system in the criminal investigation to confirm if any unauthorized traffic is entering into the network to protect data. A Raspberry Pi device has been used to deploy a honeypot as it has a constructed wireless hardware, making it more versatile and eliminating the need for an adapter, which would load up the system and jeopardize its stealthiness. For risk assessment, a custom honeypot framework is used. The functioning and application are located on the server or local host, and it can be run from the command prompt. The IP to examine the honeypot through the SSH to run the honeypot server needs to be found. The authors utilize NMAP to search for the IP within the network and then retrieve the Raspberry through PUTTY. This system protects the system from an attacker attempting to breach in through vulnerabilities [18].

This chapter is a collaboration between the members of Tencent and students at Tsinghua University. For IoT malicious forensic analysis, Tencent has created a consumer kind of IoT honeypot. It is a High-Level Interaction Honeypot (HIH) that offers more knowledge for IoT ransomware analysis and forensic analysis as compared to conventional honeypot innovation. To begin, bidirectional traffic flow will be collected, meaning that the reported information will include not only traffic aimed at attacking the machine but also traffic initiated by the infected device. The architecture of the honeypot has three layers, namely, the access layer, the storage layer, and the view layer. A load balancer was created to guarantee that ransomware in the backend execution portion has no negative effect on other entities in the same network access stage. The Raspberry Pi has been employed to be the honeypot, with the data stream being directed to Pi. Throughout the age of IoT, this model design and layout can supply every minute aspect of an intrusion, which is really useful for detecting ransomware and threats [19].

A Honeynet consists of a network environment, and the Honeynet is prepared to be assaulted after having mounted and equipped along with all the necessary equipment with the goal of preserving a safe environment for the analysis of the incidents that unfolded. It is necessary to consider the attackers' goals, strategies, and desires for the planned world by analyzing these incidents. This chapter discusses the current status of Honeynets, including frameworks, Honeynet configurations, Honeynet software, Honeynet designs, and real-world applications that are based on gathering information. A network adapter is needed for the honeypot to communicate. As a result, the honeypot is given a static IP address inside the same section. Utilizing VMware Workstation, a virtual switch is programmed to the eth0 application, which corresponds to the Honeynet's local network. Time synchronization is done with the help of the Network Time Protocol (NTP). Users are created to gain access into the honeypot. MySQL, Sebek, IPTables, and Honeywall Roo are configured and installed as well. The authors of this chapter looked at the idea of a high-level interaction in honeypot for capturing and analyzing data from assailants as well as offering details about safety risks. They can, nevertheless, be customized to collect unique data and new blackhat techniques [20].

Austin Pontén talks about how a honeypot can easily duplicate a real network with false data that an attacker will try to take advantage of, thus demonstrating how and what their objectives are. The objectives of these attackers lead to data that is crucial for building a better IDS (Intrusion Detection System), firewall, and additionally encoding the details that need to be secured. Three low-interaction-level honeypot solutions that were used in this chapter were Honeyd, Kippo, and Dionaea to understand their setup, configuration, and the capturing of data. Upon testing these three solutions, it was found that Honeyd was the best solution because of its reconfigurability, scalability, and ability to advertise services. Honeyd can duplicate an unprecedented quantity of services, including systems and specific code connected to each port. The capability to manipulate the configuration file and decide exactly how the Honeynet, or collection of honeypots, should operate is by giving the individual the power to determine what needs to be tested. The author has revealed what exactly the hackers and intruders around the world are qualified of, just in a few days, which demonstrates how important it is to create a safe and a sound network structure. A honeypot can be an asset to a network's security plan by helping in the discovery and eradication of cyberattacks [21].

In this chapter, the authors have devoted their research to execute a Raspberry Pi-based honeypot in a network that will lure intruders by modeling weaknesses and inadequate security. The honeypot will log all the hackers' events, and after data examination, it will display the type of damage and will also show improvements to be made in the security of the network. Incorporating a Raspberry Pi–Honeypot including Snort IDS is the topic of the suggested infrastructure. It will greatly reduce identification errors and data-gathering errors centered on the honeypot. As far as security tools go, they include Snort IDS, Modern Honeypot Network, Kippo, Dionaea, and Glastopf. This Raspberry Pi–Honeypot-based identification module is employed as a client–server framework. It has a centralized main server that interacts mostly with the network's various clients. The Raspberry Pi–Honeypot is a network trap that provides an easy and effective strategy for improving network security. Raspberry Pi can be used as a honeypot at a low cost and with ease of deployment, implementation, and installation. This chapter presents a novel and cost-effective information security mechanism. Throughout the phase of protecting the network, this approach blends the security software to reduce the drawbacks and optimize the security capabilities [22].

The authors of this chapter have given an extensive overview about honeypots, their software, and the methodologies used to analyze the honeypot data. The ethical and legal concerns have also been discussed. A detailed analysis has been made on the different types of honeypots such as low-level interaction honeypots, medium-level interaction honeypots, and high-level interaction honeypots. Some of the honeypot software that have been discussed are KFSensor, Honeyperl,

HoneyBow Sensor, Honeywall, SpyBye, and many more. They have even talked about various types of honeypot detection tools. A comparison study has been made between the services, applications, design, and the details of the software. The attack profile, sources, target, frequency, and evolution have been explained. The chapter concludes with the legal and ethical concerns where the authors explain how the deployment of honeypots can be a legal and ethical liability. Some of the challenges faced are entrapment, privacy, and liability [23].

A Raspberry Pi–Honeypot system including Snort IDS was incorporated. This reduced identification and data-gathering errors on the honeypot network which included Kippo, Dionaea, and Glastopf. It was employed as a client–server framework with a centralized server that interacted with the network's clients [24]. Nawrocki M. et al. gave an extensive overview about honeypots. Software like KFSensor, Honeyperl, HoneyBow Sensor, Honeywall, SpyBye, and more were analyzed using a comparison study based on services, applications, design, and the details of the software. It concluded with legal and ethical concerns on the deployment of honeypots [25].

Hardware security is responsible for safeguarding against physical and low-level threats, which cannot be effectively addressed by software security alone [26]. It assists in fulfilling legal obligations and ensures the confidentiality, genuineness, and soundness of systems. Additionally, it enables the secure execution of programs. Enterprises may enhance their security stance by using robust hardware security solutions, which offer greater reliability and resilience. Hardware security is the foundation on which software security relies, as it builds trust. Software security solutions are susceptible without a strong hardware foundation. Isolated execution environments, such as Intel SGX and ARM Trust Zone, safeguard important computations in the event that the primary operating system is corrupted. Isolated execution environments, such as Intel SGX and ARM Trust Zone, safeguard crucial computations in the event that the primary operating system is corrupted. Despite the presence of proficient assailants, equipment can still be safeguarded by anti-tampering mechanisms. The objective of these systems is to identify and respond to attempts of unauthorized physical entry. Hardware-based encryption is more effective than software-only encryption in improving data security during both transport and rest. To prevent unauthorized access or alteration, important data and cryptographic keys are held securely in Hardware Security Modules (HSMs) and Trusted Platform Modules (TPMs). Preserving the integrity of the firmware prevents unauthorized individuals from installing malicious firmware that might potentially cause damage to the machine. Verifying the genuineness of hardware components is crucial in preventing supply chain attacks that include the introduction of compromised hardware prior to deployment.

Setting up a honeypot on a Raspberry Pi necessitates the use of certain hardware. Intrusion Detection Systems (IDSs) utilize several security methods to prevent physical tampering, unauthorized access, and operational compromise [27]. To create a robust and effective honeypot using a Raspberry Pi, it is crucial to prioritize both software and hardware security measures. Physical security measures, secure installations, continuous modifications, and rigorous monitoring are necessary to protect the honeypot and the data it collects. By adhering to these procedures, you may enhance the security measures of your IDS and get a deeper understanding of potential security breaches. A hardware-secured Raspberry Pi honeypot may be used to establish a dependable intrusion detection system. To successfully identify and neutralize threats, an IDS may be constructed by giving priority to network and physical security, conducting comprehensive research, and periodically upgrading the system. This approach not only safeguards the honeypot but also exposes potential weaknesses in your network.

The detection of attack traffic from regular network traffic in Software defined Network (SDN) is a classification problem, and this kind of application trains supervised learning models to distinguish different classes of data [28]. The proposed system employs supervised Machine Learning (ML) models and trains these with labeled input data from the dataset to predict the output classification label correctly. It is simpler in SDN due to their logically centralized control logic [29] compared to traditional networks. Employing ML in Intrusion Detection Systems (IDS) for SDN is beneficial in terms of Quality of Service (QoS) and security enforcement [30–32]. ML models offer higher accuracy and lower False-Positive Rate (FAR) when applied appropriately [33, 34].

3 Problem Statement and Proposed Solution

In the growing world of technology, there is a huge demand in the field of cybersecurity as there are a lot of threats and attacks that take place every day. Hackers will always be many steps ahead of us due to their malicious intentions, so the best way to be on par with them is to use honeypots to understand their strategies and methods. The project will focus on designing, developing, and analyzing a medium-level interaction honeypot by using it in an Intrusion Detection System which will be placed on a Raspberry Pi. A Raspberry Pi is a cheap gadget that is the size of a credit card. It connects to a computer monitor or television and uses a conventional keyboard and mouse. It can perform all the functions that a desktop computer is capable of [16]. After a thorough analysis of 13 different papers based on honeypots and their applications, a medium-level interaction honeypot "Cowrie" is chosen as the honeypot for this project to create an Intrusion Detection System. The honeypot will present itself as a legitimate working network system. This honeypot will be used only for research purposes and not for production.

4 Implementation and Analysis

The work focuses on designing, developing, and analyzing a medium-level interaction honeypot while using it in an IDS, placed on a Raspberry Pi. After a thorough analysis of the 16 papers based on honeypots and their applications, honeypot

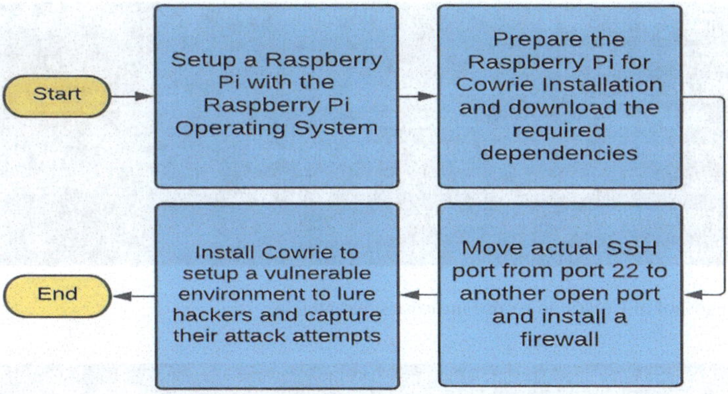

Fig. 2 Steps for the creation of a honeypot

"Cowrie" was chosen. It presented itself as a legitimate working network system and was used only for research purposes and not for production. The honeypot mimicked an SSH server, being vulnerable and easily accessible. A fake Linux Shell lured attackers and captured their steps when they attacked it. The flowchart for the creation of a honeypot is shown in Fig. 2.

Cowrie recorded the intrusions received and displayed details in its log file, as shown in Fig. 3. Malicious files downloaded by the hacker were captured as well. If the honeypot was stopped, logs would not be written. It was placed in the LAN at home, to be tested with devices within the same network. While Cowrie ran, a "ping" command was sent and the packets sent were received. The honeypot was invaded through an SSH connection that enabled a connection between two computers, and logs were recorded in real time. Changes made by the attacker were visible in the logs, such as files downloaded, the usernames and passwords used to enter the honeypot, and more.

Algorithm 1 Configuration and Working of the Cowrie Honeypot Input: Intrusion by an attacker

Output: Logs created based on intrusion attempt
Begin

1. Select a Linux shell to deploy the Cowrie honeypot
2. Update and upgrade Cowrie and Python dependencies
3. Change SSH Port from port 22 to another available port
4. Add rules for: Firewall, IP Tables and Listen Endpoints
5. Restart the System
6. Create a disabled password user and activate a virtual environment
7. Start the honeypot Intrusions are attempted and logs created in real-time
8. Stop the honeypot

End

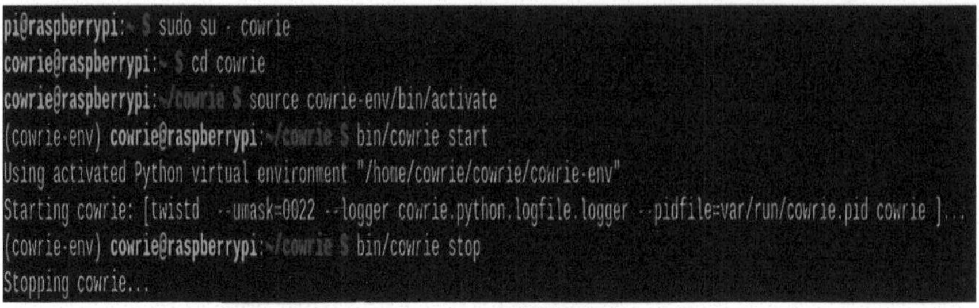

Fig. 3 Snapshot of Cowrie to record intrusions and attacks

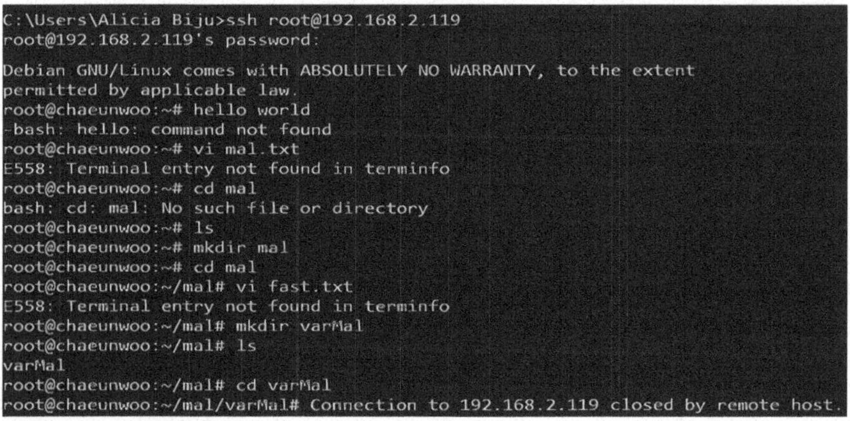

Fig. 4 SSH from Windows laptop with a valid user

5 Results and Discussion

The results and discussion are based on how the Raspberry Pi's port 22 was set as the vulnerable server, the Cowrie honeypot, and how it could record the intrusions that took place through the SSH (Secure Shell Protocol) connections from devices that were on the same network. The first intrusion was through the Windows Command Prompt, where "root" was used as the username to enter into the IP address 192.168.2.119., as illustrated in Fig. 4.

The "ssh" command by default would enter port 22 unless another port was specified. As the username "root" was saved into the "userdb.example" file which specified usernames and passwords that could gain access to the honeypot, the connection was successful as this username accepted any password. Commands to create directories and files and to change directories were run. The connection with the host was disabled when there was no activity from the intruder's side.

The second intrusion was from the Terminator, the command prompt for Kali Linux. The system ran in a virtual environment and used the username "Oracle" to enter the same IP address. This username was also saved in the file and thus gained access to the honeypot with a successful connection. Similarly, the username accepted any password and followed the same attack procedure as the first intrusion.

The last intrusion was from the Windows command prompt, where "alicia" was used as the username, as shown in Fig. 5. This username was not saved in the "userdb.example" file; hence, the connection was not successful. However, it allowed the attacker to enter a password three times to make the system look legitimate and not fake.

The honeypot logs were created in real time; for the first intrusion, it showed that it began on port 22 and then gave a message saying that the honeypot was ready to accept SSH connections. Furthermore, the logs showed that Cowrie had a new connection being made with the username "root" and password "alicia", used to setup a successful connection. Commands entered by the intruder on the Windows command prompt to modify the fake system were also shown. Similarly, a connection was created from the Kali Linux Terminator where the username "oracle" and password "f20170025Ally" were used. Directories were created, and invalid commands were run as shown in Fig. 6.

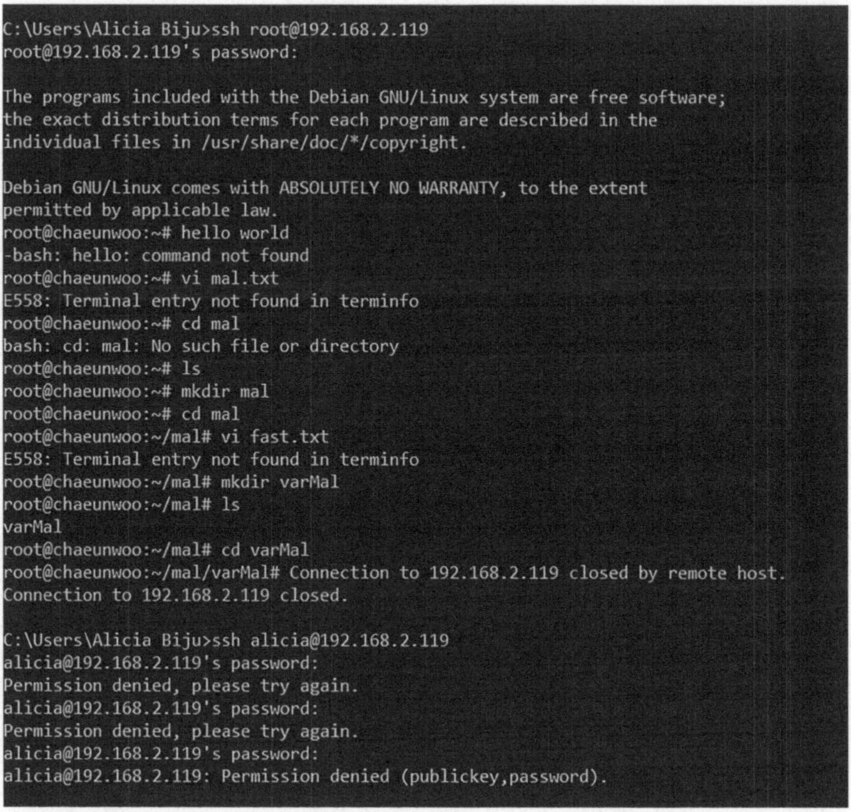

Fig. 5 SSH from Windows laptop with an invalid user

```
2021-05-19T20:34:33.486493Z [twisted.conch.ssh.session#info] Getting shell
2021-05-19T20:34:45.919695Z [HoneyPotSSHTransport,0,192.168.2.100] CMD: hello world
2021-05-19T20:34:45.930629Z [HoneyPotSSHTransport,0,192.168.2.100] Can't find command hello
2021-05-19T20:34:45.931781Z [HoneyPotSSHTransport,0,192.168.2.100] Command not found: hello world
2021-05-19T20:34:54.710769Z [HoneyPotSSHTransport,0,192.168.2.100] CMD: vi mal.txt
2021-05-19T20:34:54.722864Z [HoneyPotSSHTransport,0,192.168.2.100] Command found: vi mal.txt
2021-05-19T20:34:54.725113Z [HoneyPotSSHTransport,0,192.168.2.100] Reading txtcmd from "share/cowrie/txtcmds/usr/bin/vi"
2021-05-19T20:35:01.532778Z [HoneyPotSSHTransport,0,192.168.2.100] CMD: cd mal
2021-05-19T20:35:01.537931Z [HoneyPotSSHTransport,0,192.168.2.100] Command found: cd mal
2021-05-19T20:35:08.368785Z [HoneyPotSSHTransport,0,192.168.2.100] CMD: ls
2021-05-19T20:35:08.377772Z [HoneyPotSSHTransport,0,192.168.2.100] Command found: ls
2021-05-19T20:35:18.097036Z [HoneyPotSSHTransport,0,192.168.2.100] CMD: mkdir mal
2021-05-19T20:35:18.104143Z [HoneyPotSSHTransport,0,192.168.2.100] Command found: mkdir mal
```

Fig. 6 Cowrie logs—intruder operating through the Kali Linux terminator

```
2021-05-19T20:41:55.133189Z [cowrie.ssh.transport.HoneyPotSSHTransport#debug] starting service b'ssh-userauth'
2021-05-19T20:41:55.139627Z [cowrie.ssh.userauth.HoneyPotSSHUserAuthServer#debug] b'alicia' trying auth b'none'
2021-05-19T20:42:00.106095Z [cowrie.ssh.userauth.HoneyPotSSHUserAuthServer#debug] b'alicia' trying auth b'password'
2021-05-19T20:42:00.108709Z [HoneyPotSSHTransport,2,192.168.2.100] login attempt [b'alicia'/b'hello'] failed
2021-05-19T20:42:01.115457Z [cowrie.ssh.userauth.HoneyPotSSHUserAuthServer#debug] b'alicia' failed auth b'password'
2021-05-19T20:42:01.116814Z [cowrie.ssh.userauth.HoneyPotSSHUserAuthServer#debug] unauthorized login: ()
2021-05-19T20:42:07.175534Z [cowrie.ssh.userauth.HoneyPotSSHUserAuthServer#debug] b'alicia' trying auth b'password'
2021-05-19T20:42:07.181835Z [HoneyPotSSHTransport,2,192.168.2.100] login attempt [b'alicia'/b'sos'] failed
2021-05-19T20:42:08.189970Z [cowrie.ssh.userauth.HoneyPotSSHUserAuthServer#debug] b'alicia' failed auth b'password'
2021-05-19T20:42:08.191017Z [cowrie.ssh.userauth.HoneyPotSSHUserAuthServer#debug] unauthorized login: ()
2021-05-19T20:42:10.980799Z [cowrie.ssh.userauth.HoneyPotSSHUserAuthServer#debug] b'alicia' trying auth b'password'
2021-05-19T20:42:10.983081Z [HoneyPotSSHTransport,2,192.168.2.100] login attempt [b'alicia'/b'nooo'] failed
2021-05-19T20:42:11.989883Z [cowrie.ssh.userauth.HoneyPotSSHUserAuthServer#debug] b'alicia' failed auth b'password'
2021-05-19T20:42:11.990924Z [cowrie.ssh.userauth.HoneyPotSSHUserAuthServer#debug] unauthorized login: ()
2021-05-19T20:42:12.011888Z [cowrie.ssh.transport.HoneyPotSSHTransport#info] connection lost
2021-05-19T20:42:12.012667Z [HoneyPotSSHTransport,2,192.168.2.100] Connection lost after 16 seconds
2021-05-19T20:43:19.827521Z [-] Received SIGTERM, shutting down.
2021-05-19T20:43:19.830149Z [-] (TCP Port 22 Closed)
```

Fig. 7 Cowrie logs—Intruder operating with Windows OS

Lastly, another intrusion was made from the command prompt on the Windows OS. The connection was not established due to the fact that the username "alicia" was not stored in the "userdb.example" file. It tried to go through a second and a third attempt to login, but it did not work. A message that said the connection is lost was received. Once the honeypot was stopped, it shut down and closed the port 22 for safety purposes. It then terminated the loop, and it shut the server down as shown in Fig. 7.

6 Conclusions

The world of technology will keep moving forward faster than ever before. This means that security and safety of all data around the globe would be of utmost importance, as it could be breached at any moment. This is where the role of honeypots comes into play as it would help us understand what commands hackers use to breach into the system, malware, or malicious files they would download, and would help log other attempts at attacks as well. In this chapter, a medium-level interaction honeypot, Cowrie, was used. For the purpose of research, the SSH part of the honeypot was only investigated and port 22 was opened. The vulnerable server, honeypot, was placed on port 22 to receive traffic. It was placed inside the internal network and not next to the DMZ or external firewall for security reasons. The Cowrie honeypot was able to successfully record the intrusion attempts and provide facts such as the time and date the command was executed. It even divided the logs into distinct files based on the attacks that occurred each day. It could distinguish between valid and invalid commands, giving the user more insight into what an intruder may do in a functioning network and system. As a result, using a honeypot is a great approach to figure out how attackers operate and how to defend people from them.

References

1. Logsign. (2021). What is a Honeypot in Cybersecurity?. [online]. Available at https://www.logsign.com/blog/what-is-a-honeypot-in-cybersecurity/
2. Isong, B., Kgote, O., & Abu-Mahfouz, A. (2024). Insights into modern intrusion detection strategies for internet of things ecosystems. *Electronics, 13*, 2370. https://doi.org/10.3390/electronics13122370
3. Saheed, Y. K., & Misra, S. (2024). A voting gray wolf optimizer-based ensemble learning models for intrusion detection in the internet of things. *International Journal of Information Security, 23*, 1557–1581.
4. Javadpour, A., Ja'fari, F., Taleb, T., Shojafar, M., & Benzaïd, C. (2024). A comprehensive survey on cyber deception techniques to improve honeypot performance. *Computers & Security, 140*, 103792.
5. Nishitkumar, P., Wimmer, H., & Powell, L. (2021). Defense and analysis of hijacking user login credentials via remote code execution and raspberry pi. *Journal of Information Systems Applied Research, 14*(3), 44–50.
6. Satria, E., Huda, T. P. S., Iqbal, M., & Sarjana, F. W. (2021). The investigation on cowrie honeypot logs in establishing rule signature snort. *Journal of Physics, 644*, 012031.
7. Liu, W. K., Tan, B., Fung, J. M., Karri, R., & Chakrabarty, K. (2023). Hardware-supported patching of security bugs in hardware IP blocks. *IEEE Transactions on Computer-Aided Design of Integrated Circuits and Systems, 42*(1), 54–67.
8. Chen, F., Luo, D., Li, J., Leung, V. C. M., Li, S., & Fan, J. (2023). Arm PSA-certified IoT Chip security: A case study. Tsinghua science and technology/Tsinghua. *Science and Technology, 28*(2), 244–257.
9. Search Security. What is a honeypot? How it protects against cyber-attacks. [online] Available at https://searchsecurity.techtarget.com/definition/honey-pot
10. Mezquita, T. (2019). Honeypot – CyberHoot. [online] CyberHoot. Available at https://cyberhoot.com/cybrary/honeypot/SearchSecurity
11. Catherine, M., & Kumavat, A. (2020). Securing a small network by using raspberry pi honeypot. *International Journal of Innovative Technology and Exploring Engineering, 9*(7).
12. Vetterl, A. (2021). *Honeypots in the age of universal attacks and the Internet of Things*. https://www.cl.cam.ac.uk/techreports/UCAM-CL-TR-944.pdf
13. Bharathi, E., Keerthana, M., & Ramsundar, G. (2019). Intrusion detection using raspberry pi honeypot (snort) for network security. International research journal of. *Engineering and Technology, 6*(2), 479–481.
14. Wang, M., Santillan, J., & Fernando, A. (2018). Kuipers: ThingPot: An interactive internet-of-things honeypot. *ArXiv*. https://doi.org/10.48550/arXiv.1807.04114
15. Sharma, P., & Nagpal, B. (2022). HONEYDOS: A hybrid approach using data mining and honeypot to counter denial of service attack and malicious packets. *International Journal of Information Technology, 14*, 837–846.

16. Eliot, N., Kendall, D., & Brockway, M. (2018). A flexible laboratory environment supporting honeypot deployment for teaching real-world cybersecurity skills. *IEEE Access, 6*, 34884–34895.
17. Nkwetta, A. (2018). *Honey-system: Design, implementation and attack analysis. Computer networking and system maintenance, College of Technology*. University of Buea.
18. Anthony Raj, R., & Chayapathi, A. R. (2017). A Honeypot for a Small Network using Raspberry pi. *International Journal of Computer Engineering in Research Trends, 4*(8), 773–778.
19. Ali Ahmed, A., Wai Kit, Y., & Sallam, A. A. (2017). Raspberry Pi-based investigating model for identifying intrusion evidence. *Journal of Forensics Sciences and Criminal Investigation, 7*(3), 555715.
20. Khraisat, A., Gondal, I., Vamplew, P., et al. (2019). Survey of intrusion detection systems: Techniques, datasets and challenges. *Cybersecurity, 2*, 20.
21. Yang, J., Dang, F., Li, J., Geng, C., Liu, Z., & Zhang, Y. (2017). *An IoT honeypot device for malware forensics*. AVAR.
22. Velasco Silva, D., & Rodríguez Rafael, G. D. (2017). A review of the current state of Honeynet architectures and tools. *International Journal of Security and Networks, 12*(4), 255–272.
23. Ponten, A. (2017). *Evaluation of low-interaction honeypots on the university network* (Dissertation). Retrieved from https://urn.kb.se/resolve?urn=urn:nbn:se:lnu:diva-66885
24. Mahajan, S., Mhasku Adagale, A., & Sahare, C. (2016). Intrusion detection system using raspberry Pi honeypot in network security. *International Journal of Engineering Science and Computing, 6*(3), 41–45.
25. Nawrocki, M., Wahlisch, M., Schmidt, T. C., Keil, C., & Schonfelder, J. (2016). *A survey on honeypot software and data analysis*. Freie Universitat.
26. Hu, W., Chang, C. H., Sengupta, A., Bhunia, S., Kastner, R., & Li, H. (2021). An overview of hardware security and trust: Threats, countermeasures, and design tools. *IEEE Transactions on Computer-Aided Design of Integrated Circuits and Systems, 40*(6), 1010–1038.
27. Sivamohan, S., & Sridhar, S. S. (2023). An optimized model for network intrusion detection systems in industry 4.0 using XAI based Bi-LSTM framework. *Neural Computing & Applications, 35*(15), 11459–11475.
28. Aburomman, A. A., & Reaz, M. B. I. (2016). Survey of learning methods in intrusion detection systems. In *2016 international conference on advances in electrical, electronic and systems engineering (ICAEES)* (pp. 362–365). IEEE.
29. Sultana, N., et al. (2019). Survey on SDN based network intrusion detection system using machine learning approaches. *Peer-to-Peer Networking and Applications, 12*(2), 493–501.
30. Yang, Z., et al. (2022). A systematic literature review of methods and datasets for anomaly-based network intrusion detection. *Computers & Security, 116*, 102675.
31. Issa, M. M., Aljanabi, M., & Muhialdeen, H. M. (2024). Systematic literature review on intrusion detection systems: Research trends, algorithms, methods, datasets, and limitations. *Journal of Intelligent Systems, 33*(1).
32. Ji, I. H., Lee, J. H., Kang, M. J., Park, W. J., Jeon, S. H., & Seo, J. T. (2024). Artificial intelligence-based anomaly detection technology over encrypted traffic: A systematic literature review. *Sensors, 24*, 898.
33. Nisar, K., et al. (2020). A survey on the architecture, application, and security of software defined networking: Challenges and open issues. *Internet of Things, 12*, 100289.
34. Farooq, M. S., Riaz, S., & Alvi, A. (2023). Security and privacy issues in software-defined networking (SDN): A systematic literature review. *Electronics, 12*, 3077.

Power Up IoT: How Hardware-Assisted Blockchain Is Transforming Connectivity in Health Care

S. Mubeena, P. K. Jawahar, A. R. Abdul Rajak, and Ashutosh Mishra

1 Introduction

The integration of IoT is transforming business operations through the use of sensors, edge devices, and infrastructure. In the last 10 years, there has been a notable shift from non-IoT gadgets to IoT gadgets. It is estimated that by 2030, 75% of devices will have IoT capabilities. This can be seen in Fig. 1, which shows a comparison between non-IoT and IoT devices over the past decades. With the annual increase in interconnected devices, the volume of data managed by IoT devices has become substantial. This data moves through a sequential supply chain and is vulnerable to cyber threats. In academic literature, various networking technologies such as Wireless Sensor Networks (WSNs), Machine-to-Machine (M2M) communication, and Cyber-Physical Systems (CPS) are identified as crucial components of the broader IoT framework. Consequently, security concerns related to WSN, M2M, and CPS arise within the IoT domain, highlighting the need for robust safeguarding measures to protect the entire network infrastructure from potential security breaches facilitated by the standard IP network protocol. Without adequate security measures, IoT services are vulnerable to disruptive assaults that threaten the accuracy of data, privacy of users, and network integrity [2, 3].

According to the precedence research [4], as the industry grows, an increasing number of blockchain platforms focused on the Internet of Things are developing. The IoT landscape produces substantial volumes of data that necessitate secure

S. Mubeena · P. K. Jawahar (✉)
Department of Electronics and Communication Engineering, B.S. Abdur Rahman Crescent Institute of Science and Technology, Chennai, India
e-mail: jawahar@crescent.education

A. R. Abdul Rajak · A. Mishra
Department of Electrical and Electronics Engineering, Birla Institute of Technology and Science, Dubai Campus, United Arab Emirates

© The Author(s), under exclusive license to Springer Nature Switzerland AG 2025
A. Mishra et al. (eds.), *Hardware Security: Challenges and Solutions*,
https://doi.org/10.1007/978-3-031-81213-2_7

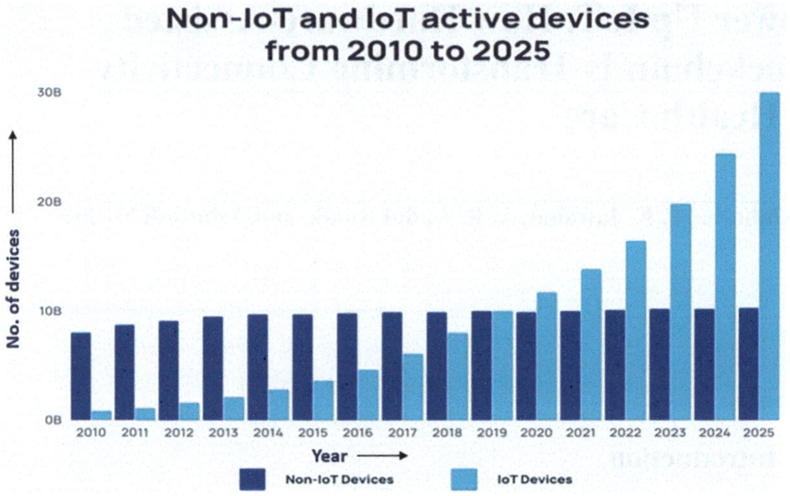

Fig. 1 Non-IoT and IoT devices from past decades [1]

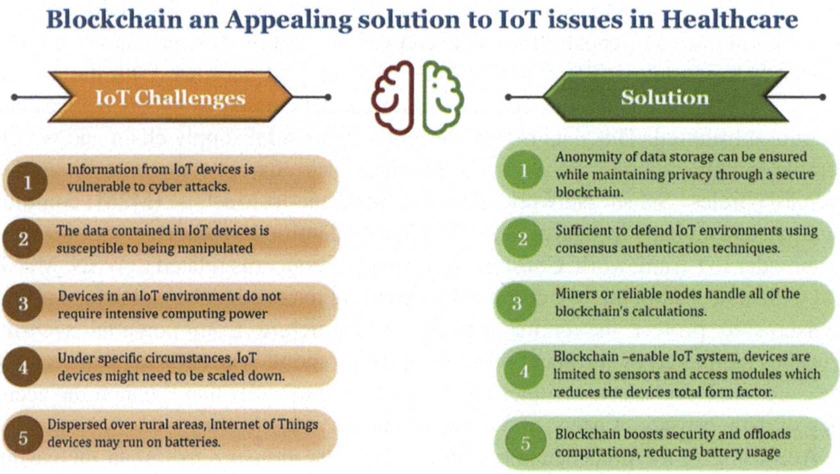

Fig. 2 Blockchain is an appealing solution to IoT issues in health care

storage, processing, and retrieval. In contrast, blockchain can be a possible solution for IoT issues in health care which is depicted in Fig. 2. Blockchain is characterized by its unchangeable and open nature, which guarantees the integrity of IoT data, thereby diminishing the likelihood of data breaches and unauthorized alterations. Blockchain, a decentralized ledger technology, facilitates machine-to-machine transactions combined with IoT. It records a series of transactions in a database, authenticated by multiple sources and stored in a shared ledger distributed across all network nodes. The convergence of IoT and blockchain offers significant benefits, such as enabling smart devices to operate autonomously without a central governing authority.

This chapter proceeds through various important sections to thoroughly investigate and support its main argument. Section 2 explores previous studies on the integration of IoT and blockchain, analyzing current solutions and discovering deficiencies in the area. Section 3 on Hardware Assistance in IoT Blockchain emphasizes how specialized hardware like ASICs, FPGAs, and secure elements play a crucial role in enhancing both the performance and security of blockchain technology. The chapter covers the fundamentals of blockchain technology and its multilayered structure. It goes on to explain the use of Truffle, Ganache, and React JS in creating and testing the system. This part consists of intricate illustrations and descriptions regarding the system's operational process and performance measurements. At last, the discussion part covers the technical difficulties and restrictions faced in the study.

2 Literature Review

Blockchain's capacity to offer secure and verifiable records renders it a suitable adjunct to IoT networks, which frequently encounter difficulties concerning data integrity, confidentiality, and security. The scholarly investigation will delve into the present status of IoT and blockchain resolutions, analyzing the amalgamation of these technologies to tackle current obstacles and develop fresh prospects. This examination will assess the merits and advantages of these incorporations, including enhanced security, data integrity, and automation facilitated by smart contracts.

The Internet of Things (IoT) faces significant challenges due to its decentralized nature. Each node in an IoT network might serve as a possible vulnerability for hostile actors to exploit to carry out cyberattacks, including Distributed Denial-of-Service (DDoS) harm. The presence of multiple compromised devices within a network can lead to rapid system failure. Additionally, the common reliance on a centralized cloud service provider introduces a single point of failure, posing a significant vulnerability that requires attention. Safeguarding data confidentiality and authentication remains a persistent and critical concern [5, 6].

Recent research studies [2] have investigated IoT applications in eHealth, smart cities and homes, supply chain management, agriculture, and the industrial sectors using blockchain technology. One major research [7] looked at the most recent blockchain breakthroughs in the Internet of Energy (IoE) sector, providing a thorough analysis of the technology's potential and uses in this field. The study focused on many applications of blockchain smart contracts in energy management, including automatic data sharing, energy transactions, and monitoring and trading of energy use on a secure blockchain peer-to-peer network.

LoRaChainCare is a novel concept for Healthcare Systems (HCS) developed in [8, 9], which combines blockchain, Fog/Edge computing, and the LoRa communication technology. A completely functional LoRa Chain Care prototype has been successfully constructed to illustrate its potential. The system consists of environmental and health sensors that are linked to an Arduino Uno board with an inbuilt ATmega328P microprocessor and LoRa shield. Furthermore, a web-based application has been developed to enable medical professionals to use blockchain services.

A secure and dependable system [10] for student attendance and registration utilizing IoT, blockchain, and QR codes was developed in the study. It offers features for student attendance, registration, and tracking transaction times. The system was implemented by creating updated Hyperledger Fabric docker images and binaries, enhancing transactions with a secure hash algorithm, and generating QR codes for transactions.

To enable decentralized decision-making processes [4] to lower maintenance costs, the study presents a decentralized controlling architecture for power system optimization procedures in local energy markets. This design is built on a decentralized genetic algorithm. A blockchain-based trade mechanism and automatic distributed resource scheduling are deployed in a simulation environment, demonstrating a revolutionary approach to energy market management. The decentralized controller framework is validated using a hardware-in-the-loop simulation platform, which shows how the suggested methodology may be applied in a real-world setting. An innovative method for researching energy management systems is the agent-based framework, which was tested on a Hardware-in-the-Loop (HIL) configuration for energy community simulations.

The research presents an IoT-BIM-BCT platform created for monitoring the asset data life cycle in offsite manufacturing. It uses a Supply Chain Management Model (SCMM-PDO) with three layers to tackle traceability problems and information exchange issues, prioritizing instant communication and compatibility. In addition, the research suggests a BCT-IoT-BIM system that utilizes Blockchain as a Service (BaaS) for managing infrastructure and smart contracts, as well as Software as a Service (SaaS) for enabling communication and transactions. Utilizing a quantitative research approach, the methodology is appropriate for pinpointing deficiencies in current knowledge and assessing the effectiveness of the SCMM-PDO model and BCT-IoT-BIM system architecture. This method allows for distinguishing facts from individual beliefs, guaranteeing that the results are consistent with empirical data. The research is focused on how the suggested model and case study can help fill the gaps identified, leading to progress in the field [11].

An innovative blockchain-powered framework for IoT services is designed to enhance security, data integrity, privacy, scalability, and interoperability. This architecture is validated by deploying a smart contract on the Polygon blockchain network within a simulated IoT scenario. The approach leverages smart contracts within the blockchain system to ensure authentication and enforce access control regulations for IoT devices and users. Transactions are secured using digital signatures and public–private key encryption with the ECDSA algorithm. Communication is divided into three stages: initialization, verification, and information transfer, each operating in distinct IoT zones managed by IoT administrators [12].

3 Hardware Assistance in IoT Blockchain

3.1 Fundamentals of Blockchain

Blockchain is a decentralized and distributed ledger, enabling information storage across multiple locations without centralized monitoring [2]. This facilitates secure and instant transfer of information and currency for individuals and businesses. In blockchain, information transfer occurs peer-to-peer, without the involvement of middlemen or intermediaries. Internet applications rely on servers owned by companies for Internet communications, while all communications pass through these servers. Similarly, we rely on banks to verify and validate our transactions when it comes to money transfers. These intermediaries, including servers, institutions, and companies, are eliminated by blockchain, presenting a new method of communication.

The first step in every blockchain transaction involves broadcasting the transaction across the entire network. After this, the legitimacy of the transaction is confirmed through validation [13]. Once validated, the transaction is included in a block. A single block is chosen from the network, and this selected block is then added to the blockchain, where it becomes a permanent fixture within the chain; this gets more elaborated in Fig. 3.

The blockchain's core technology consists of transactions, blocks, consensus, apps, and smart contracts [14]. These elements are divided into many levels, each of which is comparable to a distinct layer inside the blockchain ecosystem. Blockchain technology is generally divided into multiple layers, each crucial for maintaining the system's efficiency, security, and functionality. The infrastructure layer

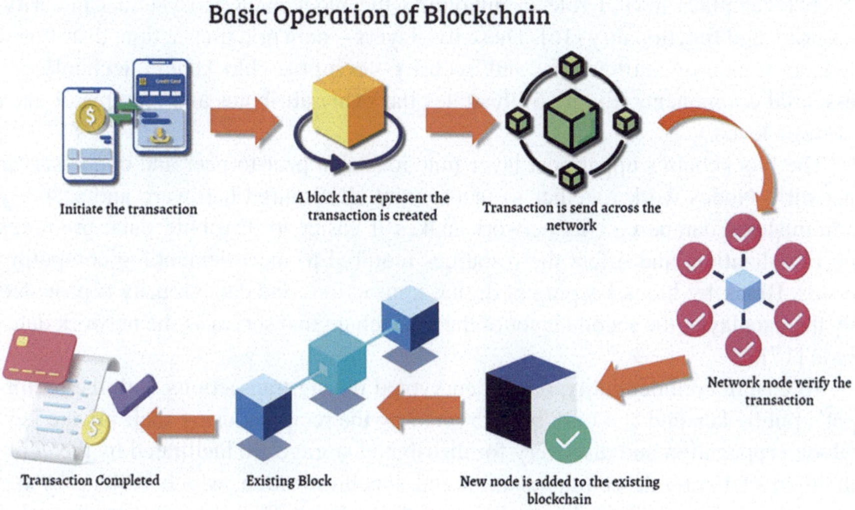

Fig. 3 Basic operation of blockchain

encompasses the hardware and networking components that support blockchain operations [38]. The data layer includes data structures and storage methods used to handle transaction records and cryptographic elements like hashes and digital signatures. The network layer manages peer-to-peer communication among nodes, facilitating consensus and transaction distribution. The consensus layer enforces protocols, such as Proof of Work (PoW) or Proof of Stake (PoS), to verify transactions and establish agreement among decentralized nodes.

Proof of Work (PoW) works by putting network members, known as miners, in a race to solve complex mathematical problems in order to authenticate transactions and produce new blocks for the blockchain. After transactions are delivered to the network, nodes combine them into a block that includes a timestamp, a link to the previous block's hash, and a nonce, which is a random integer. Miners continue to hash the block's contents with different nonces until they find a hash that matches the network's difficulty level, which needs a certain number of zeros at the beginning. Mining is a computationally intensive activity that requires a significant amount of processing labor. When a miner finds a correct hash, they send their block to the network for confirmation. Subsequent nodes validate the proof of work by hashing the block data with the nonce and verifying the outcome. When the validation of the proof of work is successful, the block is included in the blockchain, and the miner receives a combination of newly generated coins and transaction fees as a reward. Miners continue the cycle by starting work on the next block, ensuring. This is explained using the flow diagram in Fig. 4.

The incentive layer employs economic mechanisms, typically through token rewards, to encourage participation and honesty [15]. The contract layer enables the development and execution of smart contracts, allowing for automated and programmable transactions. Finally, the application layer consists of user-facing interfaces and applications, such as dApps and wallets that interact with the blockchain. Each layer plays a vital role in upholding the blockchain ecosystem's integrity, security, and functionality [16]. These five layers—network, transaction, distributed ledger, reliability, utilization, and security—comprise blockchain technology's essential components. Figure 5 illustrates the many attributes and features of each of these levels.

The blockchain's uppermost layer functions as a peer-to-peer and client–server network. Nodes work together to incorporate all required hardware and securely administer a database. This network makes it easier to distribute data, promotes decentralization, and offers the resources required to meet demanding computing needs. Block-by-block keeping of digital transactions and data security is provided by the data layer, the second layer of the blockchain that serves as the network database [17].

To ensure confidentiality, data is encrypted during transactions with the recipient's public key and can only be decrypted by the recipient using their private key. Block propagation and discovery for distributed storage are facilitated by the coordination of internode communication and synchronization, which is done by the network layer, also referred to as the propagation layer. The consensus layer is critical in blockchain because it executes protocols to assist nodes agree on adding

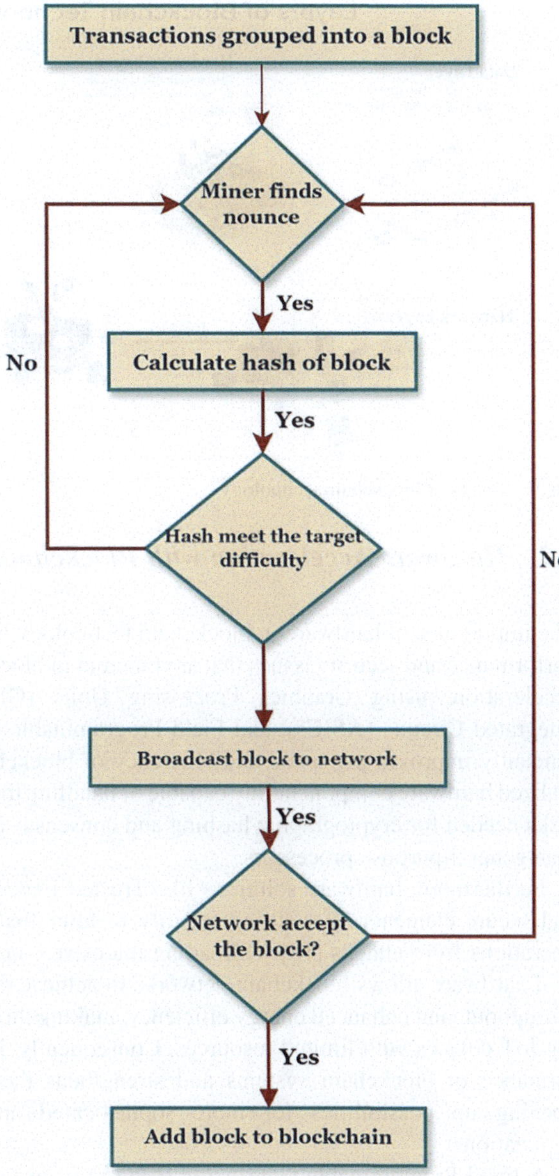

Fig. 4 Flow diagram of proof-of-work

blocks, with popular techniques including Proof of Work (PoW) and Proof of Stake (PoS). As the user-facing interface for engaging with the blockchain, for example, through digital wallet applications, the application layer also contains Smart Contracts and dApps.

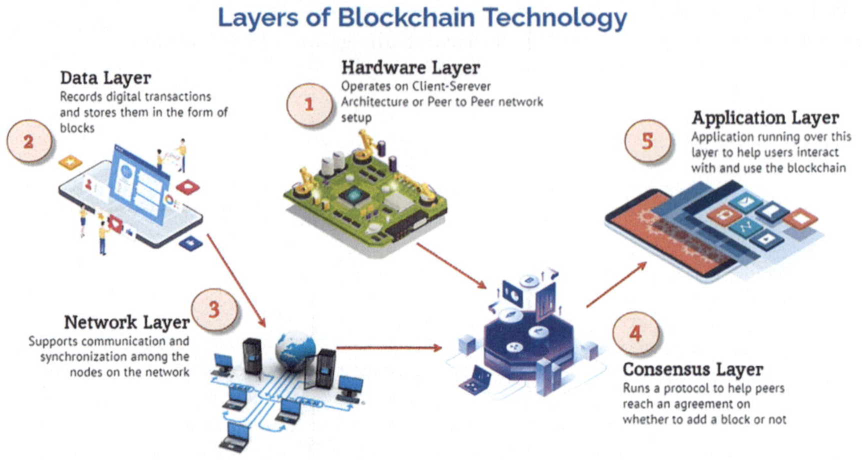

Fig. 5 Layers of blockchain technology

3.2 Hardware Acceleration with Blockchain

The importance of hardware in blockchain technology is crucial in dealing with the performance and security issues that are inherent in blockchain networks. Hardware acceleration, using Graphics Processing Units (GPUs), Application-Specific Integrated Circuits (ASICs), and Field-Programmable Gate Arrays (FPGAs), significantly improves the speed and efficiency of blockchain operations. These specialized hardware components are capable of handling the demanding computational tasks needed for cryptographic hashing and consensus algorithms more effectively than general-purpose processors.

Furthermore, hardware solutions like Trusted Execution Environments (TEEs) and secure elements offer strong security features that protect sensitive data and operations from attacks [18]. Offloading intensive computational tasks to specialized hardware allows blockchain networks to achieve decreased latency, increased throughput, and enhanced energy efficiency, making them better suited for integrating IoT devices with limited resources. Consequently, hardware improves the performance of blockchain systems and strengthens their security and scalability, opening up possibilities for more sophisticated and dependable blockchain applications.

A Field-Programmable Gate Array (FPGA) is an integrated circuit that can be altered at the hardware level and reprogrammed several times after manufacture. Leading technology firms have made significant progress in strengthening the integration of FPGA accelerators into data center servers (such as Microsoft Catapult, IBM CAPI, and Intel Xeon + FPGA) and machine learning applications [19].

A GPU-optimized array-based [17, 20] Patricia tree structure was proposed for faster blockchain search queries, leading to increased throughput when compared to CPU-based searches. Demonstrating the advantages of utilizing GPU acceleration for blockchain search and insert queries results in significant improvements in efficiency and energy usage while maintaining security, underscoring the efficacy of GPU acceleration for optimizing blockchain technology.

A novel VHDL design for PoS, PoW, and PoA consensus algorithms on the Xilinx Zynq 7000 ZC702 FPGA-based embedded platform is proposed in [21, 33]. This involves converting the POW consensus algorithm from a software version to a hardware-compatible VHDL representation, which is then packaged into a reusable IP block. The ARM processor is combined with the PoW IP block to form a unified blockchain system. The PoW IP block is first simulated using ModelSim to confirm its functionality before being deployed on the FPGA platform.

4 Proposed Design for Hardware-Assist Blockchain in Health Care

This part gives a summary of the IoT-based blockchain structure in health care and explains its key elements. Moreover, it provides more details on the workflow and offers an operational overview of the system. A block diagram of the same is depicted in Fig. 6.

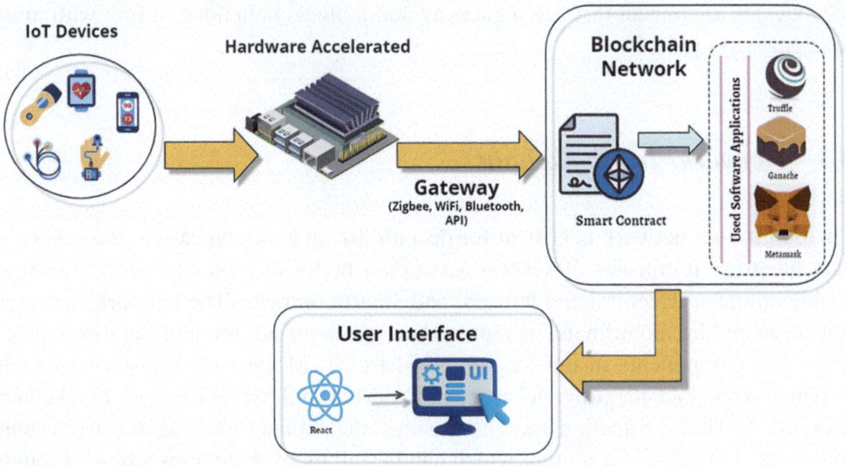

Fig. 6 Proposed design for hardware-assisted blockchain in health care

4.1 IoT Devices

IoT devices are small, resource-limited units with restrictions on processing, storage, and power capacities. Fitted with sensing and communication modules, they aid in collecting and transmitting data for purposes like environmental monitoring, smart industries, and e-health. These devices consist of microcontrollers equipped with sensors and actuators, smart wearables, as well as wireless cameras. Because of their limited resources, IoT devices do not connect to blockchain networks directly; instead, they communicate through gateway entities that have greater resources [23]. The system is compatible with different communication technologies such as WiFi, 4G, LoRaWAN, and NB-IoT. Every IoT device has a distinct identification for easier identity management in the blockchain network.

4.2 Gateway

The gateway's functions include enrolling devices, managing identities, authenticating IoT device operations, executing smart contracts, and encrypting/decrypting data. It manages the interaction between devices and the blockchain, facilitating secure IoT data communication using hardware acceleration. Intel SGX, a TEE technology, provides secure memory encryption to isolate specific application codes and data. The gateway links IoT devices to the blockchain, ensuring secure access and managing communications. It listens to smart contract events to manage secure data exchanges and supports multiple gateways with unique IDs, allowing diverse IoT devices to connect through a gateway and a blockchain node to link with multiple gateways [23].

4.3 Networks in Blockchain

The blockchain network is critical for decentralizing communication and ensuring data integrity. It consists of several networked nodes in a peer-to-peer setup that enables miners, decentralized ledgers, and smart contracts. The network, whether private or public, coordinates resource sharing, client access, and data exchange. Important components in this system consist of 1. Module for registering which provides a distinct identifier to every IoT device and symbolizes its blockchain address. 2. The verification module ensures that clients have access to system resources. 3. Module for mining which handles all blockchain transactions through miner nodes. 4. Smart contract module which deploys and oversees smart contracts along with their events. The whole process is well explained in Fig. 7 using a sequence diagram.

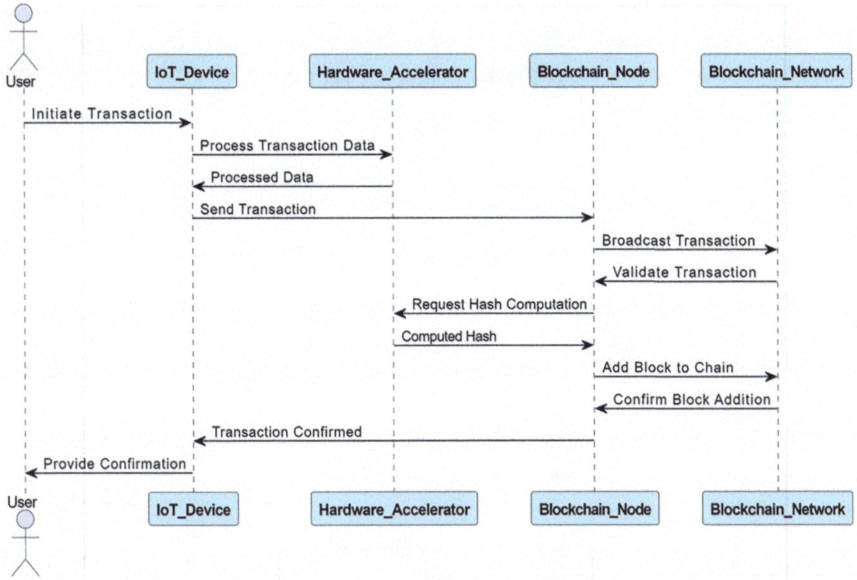

Fig. 7 Sequence diagram for hardware-assisted blockchain in health care

The suggested resolution could benefit from incorporating certain crucial enhancements. Initially, limit entry to the blockchain network to certain IP addresses or ranges, guaranteeing that only approved cloud services can link. Next, add an extra level of security by including a tool to encrypt and decrypt data leaving the blockchain network. Next, enable caching for cloud responses to minimize repeated requests; if the same request is made, the smart device can receive a previously cached response from the distributed ledger.

5 Obtained Results

Following the structure detailed in Sect. 4 and using tools like Truffle, Ganache, and React JS, the main page is displayed with a visual interface in Fig. 8.

The Ganache interface is displayed in Fig. 9 which shows a list of blocks, each featuring a single transaction. It displays the gas used by each block during the transaction. Ganache is a popular tool in the Ethereum community that enables developers to set up a virtual blockchain environment on their personal computers. It provides a private Ethereum blockchain for creating, deploying, and experimenting with smart contracts and decentralized applications (dApps) without requiring connection to the actual Ethereum network.

Block no. 23 is elaborated in Fig. 10. This block displays the gas used, gas limit, date of mining, block hash, and transaction hash. In Block 23, 712,851 gas was

Fig. 8 Patient biodata interface

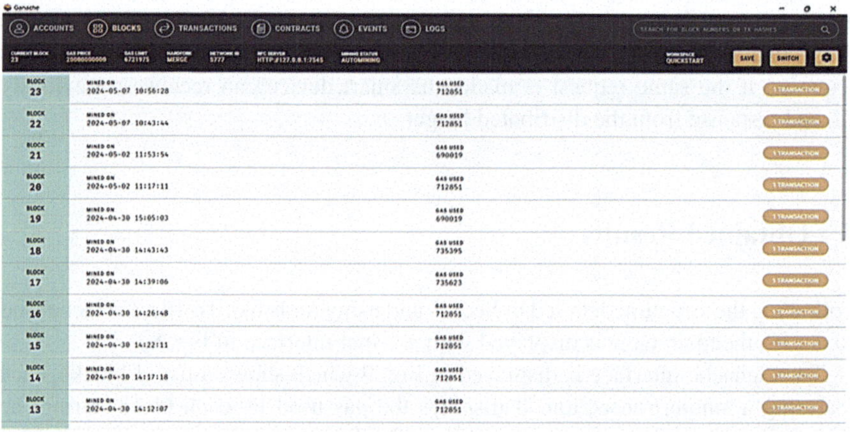

Fig. 9 Paid transaction using the Ganache interface

consumed during the transaction. In actual Ethereum transactions, around 712,851 gas equals approximately 792 Ether. This outcome was obtained through utilizing Ganache to implement and showcase the results.

Fig. 10 Obtained results from Block no. 23

Gas on the Ethereum platform measures the amount of computational work required for transactions and smart contract execution. Each network task, like sending Ether or executing a smart contract function, uses a specific quantity of gas. Users establish a gas limit to limit their spending, and any gas that is not utilized is returned [24]. If the gas consumed by a transaction goes over the set limit, the transaction will not be successful, all modifications will be reversed, and the user will still have to pay for the gas consumed.

The gas calculation involves setting a gas limit and gas price for a transaction. The gas limit represents the highest amount of gas that a user is ready to utilize, whereas the gas price (typically in gwei) indicates the Ether per gas unit that the user is willing to provide. The gas cost is determined by multiplying the gas limit and the gas price. For example, a simple Ether transaction usually requires 21,000 units of gas. Programmers utilize tools such as eth_estimateGas to predict gas usage and optimize smart contract codes to decrease gas consumption and guarantee cost-effective implementation [25].

The utilization of blockchain technology has greatly increased in the past few years, creating different platforms, each with advantages and disadvantages [26]. This conversation will center on Ethereum and Hyperledger Indy from various angles.

- **Throughput,** which is essential for handling high transaction volumes, is measured in Transactions per Second (TPS). Hyperledger Indy processes approximately 3500 transactions per second, compared to Ethereum's capacity of around 14 transactions per second.
- **Latency** refers to the duration required for a transaction to be verified. Ethereum has a longer delay than Hyperledger.
- **Expense**: Implementing smart contracts on Ethereum incurs high costs related to storing data. Hyperledger Indy cannot support smart contracts, thus eliminating costs associated. These elements affect the possibility of decentralized identification methods in extensive IoT networks.

6 Challenges and Limitations

The main problem with the existing IoT architecture is its centralized structure, leading to susceptibility to single points of failure. Blockchain solves this issue by distributing decision-making among a network of devices in public blockchain systems or one entity in private ones [27]. Three primary obstacles emerge in the integration of IoT with blockchain problems with scalability in extensive networks, challenges in maintaining transaction privacy on public blockchains, and the risk of manipulation of IoT sensor data [28–30], and this is depicted in Fig. 11. It is crucial to guarantee the integrity and security of IoT devices to ensure the safety of data transmission and transactions.

6.1 Scalability

Scalability poses a major obstacle for blockchain technologies, especially when it comes to IoT applications. In these use cases, many entities need to communicate to establish identity, grant permission, verify authenticity, exchange information, and potentially carry out smart contracts [31]. Assessments indicate that standard blockchain technologies are not suitable for handling identity and access in large networks of restricted devices because their consensus mechanisms require modification. Hence, decentralized authentication methods need to effectively

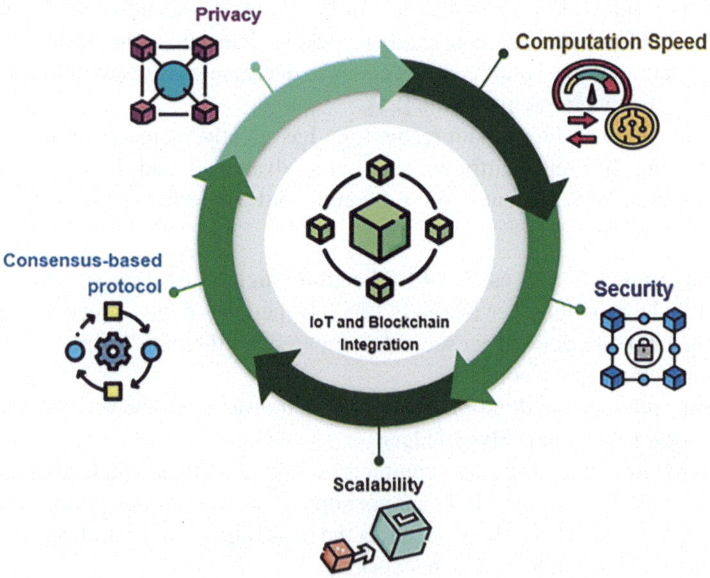

Fig. 11 Technical challenges in IoT–blockchain integration

manage substantial amounts of data in extensive networks. Methods such as sharding [32], horizontal scaling, and vertical scaling can improve scalability.

Challenges with the scalability of blockchain can be classified into three interconnected domains: speed, data storage, and expenses. The speed at which data is processed is influenced by how complicated the agreement method is and the restricted size of blocks, while the expenses associated with transactions, known as "gas" fees, also play a crucial role in the scalability of blockchain. There is a requirement for blockchain-based application solutions in different scenarios not covered in the chapter and a necessity to enhance interoperability among blockchain-based systems and applications to enable the effective deployment of smart contracts [33, 34]. Furthermore, it is important to consider the rising expenses associated with decentralized storage in blockchain networks and to create protocols that enable the modification and removal of blocks without compromising security.

Layer-Two protocols such as Lightning Network, Plasma, and Rollups have been introduced as creative methods to boost transaction speeds, lower fees, and enhance scalability by decreasing dependence on the primary blockchain [16]. These procedures make use of off-chain transactions, relying on the main chain predominantly for building trust and addressing disputes, thereby facilitating quicker and more economical transactions while upholding security and scalability. Although traditional blockchain systems face challenges handling high transaction volumes, Layer-Two solutions show potential for transforming markets and services by allowing faster transactions, reduced fees, and improved scalability.

6.2 Security

Enhancing privacy and security is achieved through decentralizing the identification process, which allows for selective information disclosure and protects against data manipulation. It is difficult to secure a large network of billions of devices, some of which have limited resources, due to the substantial computational power needed by current security protocols. Hence, an effective decentralized remedy should resist different attacks without relying on extensive computational power [35].

A protected database uses Trust Zone technology to guarantee the privacy and accuracy of electronic health records (EHRs) [36]. This involves creating a protocol for generating secure session keys using Trusted Execution Environment (TEE) to enable safe data sharing between TEE and healthcare organizations. The chapter presents important algorithms for healthcare data protection, including smart contract usage on blockchain for hashing, a secure communication protocol using TEE for session key generation, and improvements to Open Portable Trust Execution (OP-TEE) for secure database operations. Furthermore, it showcases encrypted file interfaces that are constructed based on OP-TEE's secure storage architecture.

6.3 Adaptability

Decentralized identification methods face their primary hurdle in the initial stages of development and are not being widely adopted. Although there have been continuous enhancements, none of the current frameworks have a significant number of users, which complicates the ability to forecast scalability for numerous devices. Furthermore, the absence of a uniform method for writing smart contracts leads to worries about data security. Because smart contracts are unchangeable and depend on fallible developers, any errors can have serious consequences, putting digital assets at risk of theft and misuse.

The research created an innovative approach using circular q-rung ortho-pair fuzzy sets (Cq-ROFS) and blockchain technology to tackle practical challenges in supply chain management and urban mobility [37]. The SWARA-AROMAN method was used to comprehensively assess different blockchain solutions for incorporating blockchain technology into metaverse applications for urban transportation. Furthermore, the research provides guidance and suggestions for leaders and government officials to address obstacles and optimize possibilities concerning sustainable urban transportation. Nevertheless, the method is highly dependent on input parameters, lacks flexibility, requires a lot of resources, and has limited adaptability.

6.4 Computational Speed

Utilizing blockchain for transactions with minimal data is effective, but its integration with IoT devices, which produce large amounts of data, is still difficult. An important problem is the amount of storage needed, Bitcoin and Ethereum blockchains have sizes of roughly 150 and 400 gigabytes [20], while IoT devices generate data in the zettabyte range. This information can be controlled by saving only the required data for analysis. Moreover, enhancing blockchain transaction capacity can provide stronger support for IoT information.

The research conducts an in-depth analysis of GPU [39] information system architecture and its impact on blockchain processing efficiency, setting energy efficiency standards and performance benchmarks. Factors like GPU core speed, temperature, core memory speed, ETASH algorithms, fan speed, and energy consumption are essential in deciding the computational and energy efficiency in blockchain mining. Improved core memory speed enhances both performance and energy efficiency, while elevated core temperature hinders these measures.

The ETASH algorithm is recognized as a crucial indicator of performance. The proof-of-work algorithm used to mine Ethereum and other ETH-based coins is called Ethash. It is used by Ethereum, Ethereum Classic, and numerous other digital currencies in their networks. Ethash, a modified version of the Dagger-Hashimoto algorithm, contains unique properties that make it ASIC-resistant and easily verifiable.

The methodology of the study involves creating a uniform data collection protocol, arranging and preparing the data, and using multilinear regression analysis to investigate the correlation between GPU configurations and energy efficiency.

7 Conclusions

In today's digital revolution, the demand for IoT is rapidly growing. IoT's primary challenges stem from its centralized structure, which can be addressed by leveraging blockchain's decentralized architecture to enhance scalability and security. Blockchain's distributed ledger technology mitigates single points of failure, secures transactions, and ensures robust data protection. This chapter explores the integration of IoT with blockchain, emphasizing how hardware-supported blockchain can improve security, scalability, and efficiency in IoT systems. Various specialized hardware technologies like ASICs, FPGAs, and secure elements were discussed.

The proposed design significantly improves latency, energy efficiency, and throughput by integrating hardware components with IoT devices. Although progress has been made, technical challenges and limitations remain, highlighting areas for future improvement. Overall, this research underscores the transformative potential of hardware-supported blockchain for IoT, promoting enhanced data management and increased security and efficiency across diverse applications.

References

1. Non-IoT and IoT active devices. https://iot-analytics.com/state-of-the-iot-2020-12-billion-iot-connections-surpassing-non-iot-for-the-first-time/
2. Md Ashraf, U., et al. (2021). A survey on the adoption of blockchain in IoT: Challenges and solutions. *Blockchain: Research and Applications, 2*(2), 100006.
3. Huan, N. T. Y., & Zukarnain, Z. A. (2024). A survey on addressing IoT security issues by embedding blockchain technology solutions: Review, attacks, current trends, and applications. *IEEE Access, 99*, 1–1.
4. Galicia, M., Mureddu, M., Ghiani, E., & Pilo, F. (2022). Blockchain-based hardware-in-the-loop simulation of a decentralized controller for local energy communities. *Energies, 15*(20), 7623.
5. Panarello, A., Tapas, N., Merlino, G., Longo, F., & Puliafito, A. (2018). Blockchain and IoT integration: A systematic survey. *Sensors, 18*(8), 2575.
6. Rejeb, A., Rejeb, K., Appolloni, A., Jagtap, S., Iranmanesh, M., Alghamdi, S., Alhasawi, Y., & Kayikci, Y. (2024). Unleashing the power of the internet of things and blockchain: A comprehensive analysis and future directions. *Internet of Things and Cyber-Physical Systems, 4*, 1–18.
7. Miglani, A., Kumar, N., Chamola, V., & Zeadally, S. (2020). Blockchain for Internet of Energy management: Review, solutions, and challenges. *Computer Communications, 151*, 395–418.

8. Queralta, J. P., Gia, T. N., Tenhunen, H., & Westerlund, T. (2019). Edge-AI in LoRa-based health monitoring: Fall detection system with fog computing and LSTM recurrent neural networks. In *2019 42nd International conference on Telecommunications and Signal Processing (TSP)* (pp. 601–604). IEEE.
9. Dammak, B., et al. (2022). Lorachaincare: An IoT architecture integrating blockchain and Lora network for personal health care data monitoring. *Sensors, 22*(4), 1497.
10. Abas, S. U., Duran, F., & Tekerek, A. (2023). A Raspberry Pi-based blockchain application on IoT security. *Expert Systems with Applications, 229*, 120486.
11. Brandín, R., & Abrishami, S. (2024). IoT-BIM and blockchain integration for enhanced data traceability in offsite manufacturing. *Automation in Construction, 159*, 105266. (2024).
12. Deep, A., Perrusquía, A., Aljaburi, L., Al-Rubaye, S., & Guo, W. (2024). A novel distributed authentication of blockchain technology integration in IoT services. *IEEE Access, 12*, 9550.
13. Huang, J., Kong, L., Chen, G., Wu, M. Y., Liu, X., & Zeng, P. (2019). Towards secure industrial IoT: Blockchain system with credit-based consensus mechanism. *IEEE Transactions on Industrial Informatics, 15*(6), 3680–3689.
14. Ahmad, I., Alqarni, M. A., Almazroi, A. A., & Alam, L. (2021). Real estate *management* via a decentralized blockchain platform. *Computers, Materials and Continua, 66*(2), 1813–1822.
15. Bellaj, B., Ouaddah, A., Bertin, E., Crespi, N., & Mezrioui, A. (2024). Drawing the boundaries between blockchain and blockchain-like systems: A comprehensive survey on distributed ledger technologies. *Proceedings of the IEEE, 112*(3), 247–299.
16. Rebello, G. A. F., Camilo, G. F., de Souza, L. A. C., Potop-Butucaru, M., de Amorim, M. D., Campista, M. E. M., & Costa, L. H. M. (2024). A survey on blockchain scalability: From hardware to layer-two protocols. *IEEE Communications Surveys & Tutorials., 26*, 2411–2458.
17. Tabatabaei, M. H., Vitenberg, R., & Veeraragavan, N. R. (2023). Understanding blockchain: Definitions, architecture, design, and system comparison. *Computer Science Review, 50*, 100575.
18. Mahony, A. O., & Popovici, E. (2019, June). A systematic review of blockchain hardware acceleration architectures. In *2019 30th Irish Signals and Systems Conference (ISSC)* (pp. 1–6). IEEE.
19. Turan, F., & Verbauwhede, I. (2020). Trust in FPGA-accelerated cloud computing. *ACM Computing Surveys (CSUR), 53*(6), 1–28.
20. Ifrim, R., Loghin, D., & Popescu, D. (2023, June). Baldur: A hybrid blockchain database with FPGA or GPU acceleration. In *Proceedings of the 1st workshop on verifiable database systems* (pp. 19–27).
21. Kari, J., Frikha, T., Hamdi, M., & Hamam, H. (2024). Enhancing blockchain consensus with FPGA: Accelerating implementation for efficiency. *IEEE Access, 12*, 44773–44785.
22. Chatterjee, S., Bhattacharya, P., & De, D. (2024). Light dew: Lightweight blockchain assisted dew computing framework for smart assisted living. *IEEE Transactions on Consumer Electronics, 70*, 4804.
23. Fernando, E. (2019). Blockchain technology implementation in Raspberry Pi for private network. In *2019 international conference on Sustainable Information Engineering and Technology (SIET)* (pp. 154–158). IEEE.
24. Arif, Y. M., Imron, M. K. A., Hanani, A., Nurhayati, H., Karami, A. F., & Hariyadi, M. A. (2024). Blockchain as a learning tool: Analyzing transaction speeds at various gas limits in computer assembly simulations. *International Journal of Intelligent Engineering and Systems, 17*(2), 728–739.
25. Ferenczi, A., & Bădică, C. (2024). Prediction of Ethereum gas prices using Deep AR and probabilistic forecasting. *Journal of Information and Telecommunication, 8*(1), 18–32.
26. Hosseini, S. M., Ferrira, J., & Bartolomeu, P. C. (2023). Blockchain-based decentralized identification in IoT: An overview of existing frameworks and their limitations. *Electronics, 12*(6), 1283.
27. Ray, P. P., Dash, D., Salah, K., & Kumar, N. (2020). Blockchain for IoT-based healthcare: Background, consensus, platforms, and use cases. *IEEE Systems Journal, 15*(1), 85–94.

28. Bary, T. A. A. A. A., Elomda, B. M., & Hassan, H. A. (2024). Multiple layer public blockchain approach for Internet of Things (IoT) systems. *IEEE Access, 12*, 56431.
29. Šarac, M., Pavlović, N., Adamović, S., Saračević, M., & Radovanović, D. (2023). Blockchain implementation for IoT devices: Blockchain of things. In *Blockchain technology solutions for the security of IoT-based healthcare systems* (pp. 151–165). Academic.
30. Biswas, K., Chowdhury, M. J. M., & Usman, M. (2024). Blockchain of things: Benefits, challenges, and future directions. *Sensors, 24*(3), 934.
31. Albulayhi, A. S., & Alsukayti, I. S. (2023). A blockchain-centric IoT architecture for effective smart contract-based management of IoT data communications. *Electronics, 12*(12), 2564.
32. Al-Mutar, F. H. N., Al-Khazaali, A. A. T., & Hataf, B. A. (2024). Scalability of blockchain: Review of cross-sharding with high communication overhead. In *BIO web of conferences* (Vol. 97, p. 00075). EDP Sciences.
33. Alghamdi, T. A., Khalid, R., & Javaid, N. (2024). A survey of blockchain-based systems: Scalability issues and solutions, applications and future challenges. *IEEE Access, 12*, 79626.
34. Jain, A. K., Gupta, N., & Gupta, B. B. (2024). A survey on scalable consensus algorithms for blockchain technology. *Cyber Security and Applications, 3*, 100065.
35. Bouraga, S. (2021). A taxonomy of blockchain consensus protocols: A survey and classification framework. *Expert Systems with Applications, 168*, 114384.
36. Li, J., Luo, X., & Lei, H. (2024). TrustHealth: Enhancing eHealth Security with Blockchain and Trusted Execution Environments. *Electronics, 13*(12), 2425.
37. Zhou, S., Yu, L., Wang, Y., Dhahbi, S., Berrima, M., & Anjum, M. (2024). Adaptive solutions for metaverse urban mobility through decision-making and blockchain. *Alexandria Engineering Journal, 107*, 1–14.
38. Mohammad, A. S., Brohi, M. N., & Khan, I. A. (2021). Integration of IoT and blockchain. *Technium: Romanian Journal of Applied Sciences and Technology, 3*(8), 32–41.
39. Rukhiran, M., Boonsong, S., & Netinant, P. (2024). Sustainable optimizing performance and energy efficiency in proof of work blockchain: A multilinear regression approach. *Sustainability, 16*(4), 1519.

Physical Layer Security for Future Wireless Communication Systems

K. Gunaseelan, R. Dhanusuya, A. R. Abdul Rajak, and Ashutosh Mishra

1 Introduction

With the expansion of wireless networks and creation of new mobile applications (M-App), the number of active mobile subscribers is rising quickly. While superposition can cause numerous signals to overlap at the receiver, the broadcast nature makes it challenging to conceal transmitted signals from unintentional recipients. This results in the emergence of network vulnerabilities including information jamming, hacking, malicious attacks, and eavesdropping. Security is therefore an inherent component of any communication technology that is used to transfer sensitive data. Wireless communication and PLS are inherently related because PLS directly takes advantage of the unique features of wireless channels to improve security. To achieve absolute secrecy, it investigates the properties of the channel connecting the authorized sender to the intended recipient of the wireless communication system [1, 2]. Relying on concepts like the wiretap channel and secrecy capacity, PLS approaches provide robust protection against physical layer attacks like eavesdropping. When resources are limited and typical cryptography techniques are impracticable, these techniques are especially useful. Hence, PLS is positioned to be a key component of secure wireless communication in the future, with applications ranging from smart grids and military communications to IoT and automotive networks.

K. Gunaseelan (✉) · R. Dhanusuya
Department of Electronics and Communication, College of Engineering Guindy, Anna University, Chennai, India
e-mail: guna_2012@annauniv.edu

A. R. A. Rajak · A. Mishra
Department of Electrical and Electronics Engineering, Birla Institute of Technology and Science, Pilani-Dubai Campus, Dubai, UAE
e-mail: abdulrazak@dubai.bits-pilani.ac.in; ashutosh@dubai.bits-pilani.ac.in

© The Author(s), under exclusive license to Springer Nature Switzerland AG 2025

A. Mishra et al. (eds.), *Hardware Security: Challenges and Solutions*,
https://doi.org/10.1007/978-3-031-81213-2_8

2 Security Challenges of Wireless Communications

The term "wireless communications security" refers to any security measures that keep data integrity and confidentiality intact while also guarding against unwanted access or attacks to data transferred over wireless networks. The security requirements of the wireless communication system can be broadly classified into six categories, i.e., authenticity, confidentiality, integrity, access control (privacy), nonrepudiation, and availability. The security requirements and the corresponding cryptographic solutions are described in this section.

- **Authenticity:** In order to guarantee that only authorized users are able to access the network, authentication verifies the identities of people and devices participating in wireless communication. Digital Signatures, Public Key Infrastructure (PKI) framework, and Remote Authentication Dial-In User Service (RADIUS) protocol are the examples of authentication techniques.
- **Confidentiality:** Information sent over a wireless channel is protected so that only authorized users can access it. It keeps the data from being intercepted and interpreted by unauthorized users. Using techniques like encryption, frequency hopping, and beamforming ensures confidentiality of information in wireless networks. Advanced Encryption Standard (AES) is the commonly used encryption technique to provide confidentiality.
- **Integrity:** Integrity assures that no tampering or alteration has occurred to the data being transmitted over a wireless channel. It ensures that the data transmitted and received are the same. By using Hash functions (e.g., Secure Hash Algorithm, SHA-512), Message Authentication Codes (MACs), and Digital Signatures (DSs), integrity is provided.
- **Privacy:** Ensuring privacy entails making sure that user actions, communication content, and personal information are kept private and not accessible to unauthorized users. It includes several dimensions, like data privacy, location privacy, and contextual privacy. End-to-End Encryption (E2EE) and Transport Layer Security (TLS) are some of the key privacy techniques.
- **Nonrepudiation:** Nonrepudiation makes sure that neither the sender nor the recipient of a message can retract their actions. It offers evidence of the data's transmission and origin. Among the techniques for implementing nonrepudiation are RSA (Rivest–Shamir–Adleman) algorithm and blockchain.
- **Availability:** Availability assures that, when required, wireless communication services will be accessible and operational. It guards against interferences that could make the network unusable, including Denial of Service (DoS) attacks. Implementing redundancy, antijamming techniques like DSSS (Direct Sequence Spread Spectrum), and load balancing ensures availability.

To ensure the security of communication systems, one or more of these security mechanisms must be applied. Robust, efficient, and effective security measures are needed to overcome the aforementioned security challenges in wireless communications.

3 OSI Model

Any data transmitted across a network follows the Open System Interconnect (OSI) model, i.e., the OSI model specifies how data is transmitted and received across a network. In 1978, the International Standards Organization (ISO) introduced it as a preliminary measure toward mandating globally standardized protocols for transmitting data. It describes a seven-layer hierarchical architecture that conceptually separates networking functions required to enable system-to-system connections. As a result, each layer can be regarded as a separate module. Every layer performs different tasks and has corresponding security vulnerability.

Thus, comprehensive protection is ensured by putting security controls and protocols in place at every layer. This includes implementing access controls, firewalls, intrusion detection systems, encryption, and other layer-appropriate security measures. Figure 1. shows the different layers, their security vulnerabilities, and the corresponding mitigation techniques to overcome these issues.

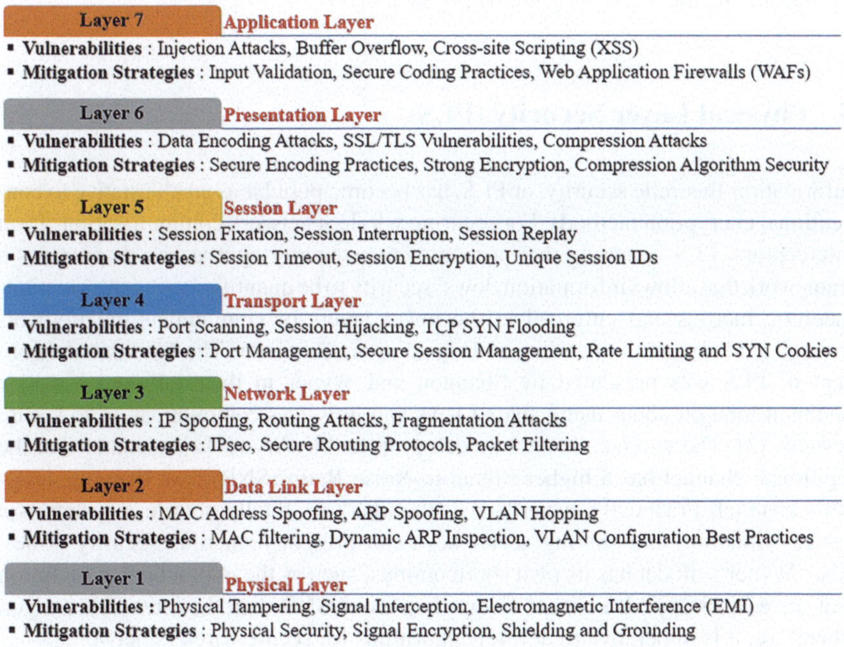

Fig. 1 Layers of OSI model security vulnerabilities and mitigation strategies

4 Security Limitations of Traditional Cryptographic Approaches

Traditionally, issues concerning the authenticity, secrecy, and privacy of data transmission are handled by the upper layers of the OSI model. Conventional cryptographic algorithms which include symmetric encryption, public key, and secret key distribution provide security based on mathematical problems that are difficult to break by conventional computers. The RSA technique, for instance, provides strong security strength in digital signatures since it is impossible for a traditional computer to factorize very large integers efficiently. This kind of traditional cryptography still makes up the majority of the encryption utilized in modern, cutting-edge systems. Inevitably, the security strength of the implemented cryptographic primitives (well-established base algorithms) can be compromised if the underlying mathematical problem can be solved efficiently by quantum computers. Furthermore, due to the intricate/complex infrastructure requirements for key management, traditional schemes will not sustain long. All these conventional cryptographic algorithms operate independent from the physical layer.

5 Physical Layer Security (PLS)

Information-theoretic security, or PLS, has become popular as an alternative to conventional encryption methods. For securing wireless network communications from interceptors, PLS is among the more attractive security approaches. It provides a framework that allows information flows' security to be quantified using information-theoretic metrics and enforced at the lower layers of communication protocols through a combination of coding and signaling mechanisms. The fundamental concept of PLS was presented by Shannon and Wyner in their influential works. Shannon thought about using shared keys and noiseless communication to secure systems [3]. The wiretap channel model, proposed by Wyner [4], assumes that the legitimate channel has a higher Signal-to-Noise Ratio (SNR) than the eavesdropper's channel. Practically, nevertheless, the idea of a totally noiseless system is a perfect situation, and sharing secret keys can present numerous security issues. Also, Wyner's model has its own shortcomings, such as the impractical assumption that the eavesdropper's channel always has lower SNR than the intended receiver. Therefore, it is imperative to develop algorithms for secure wireless networks.

The underlying idea of PLS is to exploit the inherent unpredictability of noise and communication channels to restrict the amount of data that an unauthorized user may obtain at the "bit" level.

As previously mentioned, PLS makes use of methods for signal processing and design to ensure keyless secure data transmission by taking advantage of the inherent features of wireless channels, such as interference, noise, and fading, to decrease the eavesdropper's signal reception and improve the receiver's signal quality [5].

Most crucially, the achieved security can be precisely quantified, and no limitations are assumed about the eavesdropper's processing power or knowledge of network parameters. Further, when compared to encryption-based methods, the PLS approaches incur very small overhead to relatively complete the signal-processing algorithms. Physical layer techniques also have the potential to supplement current security measures because they can function independently of higher levels. It is anticipated that using such a multilayered strategy will greatly improve the security of modern wired and wireless data networks.

5.1 Claude Shannon: The Father of Information Theory

Shannon established the concepts that would form the cornerstone of secure communication in his groundbreaking article. In general, the definition of a secrecy system is a group of transformations from one space (the set of possible communications) to another (the set of possible cryptograms). Figure 2. depicts a schematic representation of a general secrecy system.

A message source and a key source are the two information sources at the transmitting end. The key source generates a specific key from the set of possible keys within the system. This key is sent to the recipient by a method that is purportedly uninterceptable. The message source generates a message (the "clear"), which is then transferred to the recipient via a potentially interceptable channel (like radio) after being encrypted. The message is recovered at the receiving end by combining the decipherer's key and cryptogram.

An operational function involving the message and key is carried out by the encipherer. Given a message M, a key K, the enciphered message, or ciphertext, C, is represented as

$$C = f(M,K) \tag{1}$$

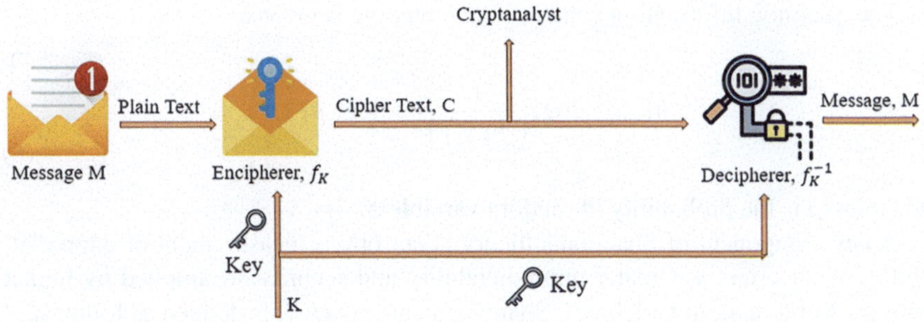

Fig. 2 Conventional secrecy system

Nevertheless, it is better to consider C as a (one parameter) family of operations or transformations rather than as a function of two variables.

$$C = f_K M. \tag{2}$$

Cipher text, C is the result of applying transformation f_K on message M. The index K is associated with the specific key being used.

Recovering M at the receiving end is possible with the knowledge of C and K. Therefore, each of the family's transformations must have a distinct inverse f_K^{-1} such that $f_K f_K^{-1} = I$, the identity transformation. Hence,

$$M = f_K^{-1} C \tag{3}$$

In any case, for any C that can be produced from a M using key K, this inverse must exist uniquely.

5.2 *Information-Theoretic Security*

According to Shannon, the term "channel capacity" refers to the highest rate at which information can be transmitted across a communication channel with minimal error probability. Given an additive white Gaussian noise (AWGN) channel, its capacity is given by

$$C = B \log_2 \left(1 + SNR\right) \tag{4}$$

where SNR is the signal-to-noise ratio and B is the bandwidth of the channel (fixed quantity) in Hertz (Hz). The SNR is mathematically expressed as

$$SNR = \frac{P}{\sigma^2} \tag{5}$$

where P and σ^2 denote the signal power and noise power, respectively.

The Shannon information content of an outcome is given as

$$h(x_i) = \log_2 \left(\frac{1}{p(x_i)} \right) = -\log_2 p(x_i) \tag{6}$$

where $p(x_i)$ is the probability of random variable X.

A key component of Shannon's theory is entropy, a measurement of unpredictability or uncertainty. Greater unpredictability and security are implied by higher entropy. For a random variable X, Shannon's entropy $H(X)$ is defined as follows:

$$H(X) = -\sum_i p(x_i)\log_2 p(x_i) \tag{7}$$

The mutual information, $I(X;Y)$, defines the amount of information X conveys to Y:

$$I(X;Y) = H(X) - H(X/Y) \tag{8}$$

So, we can infer from the above equation that the amount of information sent to Bob by Alice is given by

$$I(X;Y) = H(X) - H(X/Y) \tag{9}$$

whereas the amount of information shared between Alice and Eve is given by

$$I(X;Z) = H(X) - H(X/Z) \tag{10}$$

where the bits transmitted by Alice is represented by random variable X. Bob's received bits are denoted by Y, and the bits received by Eve are indicated by Z.

In information-theoretic secrecy, to ensure secure communication on wireless channel, it is important to maximize the mutual information that can be achieved, i.e., the secrecy rate $C_S = I(X;Y) - I(X;Z)$ by optimizing the input distribution $p(X)$. The secrecy rate of the PLS system is defined as the difference between the capacities of Bob's and Eve's channels.

Perfect Secrecy

We use Shannon's wiretap channel model, depicted in Fig. 3, to define perfect secrecy. In this concept, Alice and Bob are the sender and recipient, respectively, while Eve is the eavesdropper who records a degraded version of the communication.

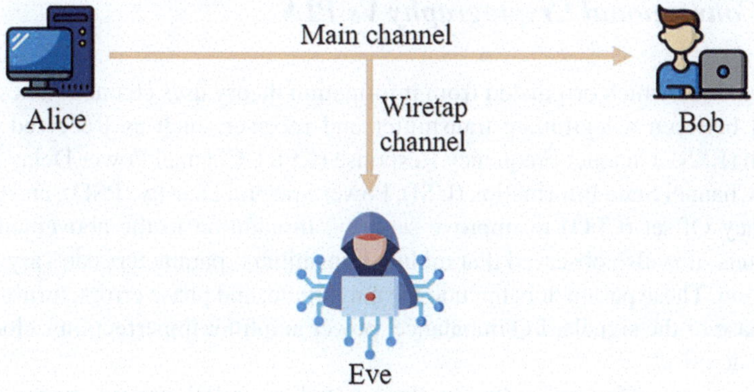

Fig. 3 Wiretap channel

If X is the transmitted signal, Y and Z are Bob's and Eve's received signals, respectively, the wiretap channel can be expressed as

$$Y = X + N_B \tag{11}$$

$$Z = X + N_E \tag{12}$$

The noise elements in Bob's and Eve's channels are N_B and N_E, respectively.

Perfect secrecy was characterized by Shannon as the state in which, regardless of the available computational capabilities, an adversary cannot deduce the original message from the ciphertext as it gives no information about the plaintext. This is expressed mathematically as

$$H(X/Z) = H(X) \tag{13}$$

Therefore, $I(X; Z) = 0$.

Aaron Wyner built on Shannon's work in the 1970s by concentrating on workable secure communication over noisy channels. His work is essential to comprehending secure transmission in practical scenarios. Wyner proposed the modified wiretap channel with encoder and decoder and demonstrated that by properly encoding the message, Alice can guarantee that Bob can decode it correctly while Eve can still not figure out the message from the observations.

5.3 Conventional Cryptography Vs PLS

PLS and conventional cryptography are compared according to several characteristics/various parameters in Table 1.

5.4 Conventional Cryptography Vs PLS

PLS technique which originated from information theory uses characteristics of the channel between a legitimate transmitter and receiver, such as Received Signal Strength (RSS), Channel Frequency Response (CFR), Channel Power Delay Profile (PDP), Channel State Information (CSI), Power Spectral Density (PSD), and Carrier Frequency Offset (CFO) to improve security. In addition to the abovementioned parameters, it is also observed that mobile transmitters' parameters can vary during fabrication. These parameters include the magnitude, and phase errors, turn-on transient phase of the signals, I/Q imbalance, power amplifier imperfections, clock offset, and dc offset.

Based on the information theory, there is a direct relationship between mutual information and channel security [6]. In order to derive this relation, consider a

Table 1 Conventional cryptography vs PLS

Parameter	Conventional cryptography	PLS
Layer of operation	Higher layers of the OSI model (application, transport)	Physical layer of the OSI model
Security basis	Computational hardness assumptions	Physical properties of the communication medium
Primary techniques	Symmetric and asymmetric encryption, hash functions, digital signatures	Artificial noise generation, beamforming, cooperative jamming, secret key generation from channel characteristics
Key management	Requires secure key distribution and management	Can derive keys from channel characteristics, reducing key management complexity
Resilience to attacks	Based on mathematical difficulty of breaking algorithms	Based on unpredictability and variability of wireless channels
Performance impact	Higher computational overhead	Minimal impact on throughput but requires additional signal processing
Scalability	Scalable with effective key management	Scalability can be challenging due to hardware and channel estimation requirements
Implementation complexity	Generally lower hardware complexity, higher computational complexity	Higher hardware complexity (e.g., MIMO systems), advanced signal processing required

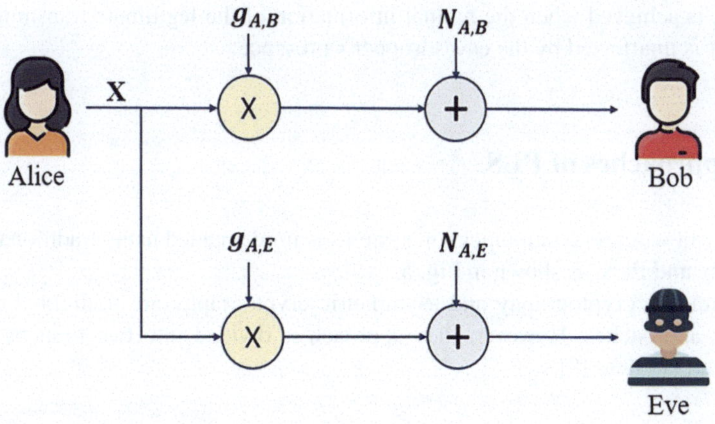

Fig. 4 Basic WPLS system model

basic WPLS system shown in Fig. 4, with a single transmitter, receiver, and an eavesdropper over the Rayleigh fading channel [7].

The received signals at Bob and Eve are given by

$$Y_{A,B} = \sqrt{P_S}\, g_{A,B} X + N_{A,B} \tag{14}$$

$$Y_{A,E} = \sqrt{P_S} g_{A,E} X + N_{A,E} \tag{15}$$

$$g_{A,B} = \left| h_{A,B} \right|^2 / d_{A,B}^{-n} \tag{16}$$

where n is the path loss exponent, P_S is the signal power; $g_{A,\,B}$ and $d_{A,\,B}$ represent the channel gain and distance between Alice and Bob, respectively.

The SNR for Alice to Bob ($\gamma_{A,\,B}$) and Alice to Eve ($\gamma_{A,\,E}$), respectively, are

$$\gamma_{A,B} = P_S \left| h_{A,B} \right|^2 / N_{A,B} \tag{17}$$

$$\gamma_{A,E} = P_S \left| h_{A,E} \right|^2 / N_{A,E} \tag{18}$$

Using the above equations, the mutual information for the legitimate and wiretap links are written as

$$I(A;B) = \log_2 \left(1 + \gamma_{A,B}\right) \tag{19}$$

$$I(A;E) = \log_2 \left(1 + \gamma_{A,E}\right) \tag{20}$$

Comparing with the Wyner's wiretap model, according to Shannon's theory, perfect secrecy is achieved when the mutual information of the legitimate transmitter and receiver is unaffected by the eavesdropper's presence.

6 Approaches of PLS

Security in wireless communication systems is implemented using traditional cryptography and PLS, as shown in Fig. 5.

Symmetric cryptography and asymmetric cryptography are traditional cryptographic approaches. However, PLS approach is divided into two branches: key-based and key-less [8].

6.1 Key-Based Approach

With key-based PLS, the authorized transmitter and receiver obtain keys from a common source. The channel itself is the source in wireless communication systems since its frequency, time, and space can change arbitrarily.

In this approach, temporal variation and reciprocity properties of the wireless channel are exploited in the generation of shared secret keys. Hence, PLS schemes

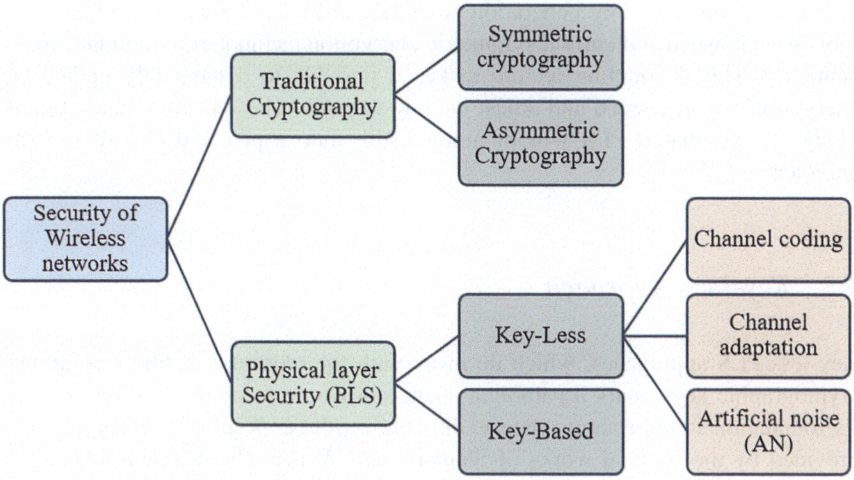

Fig. 5 Wireless network security

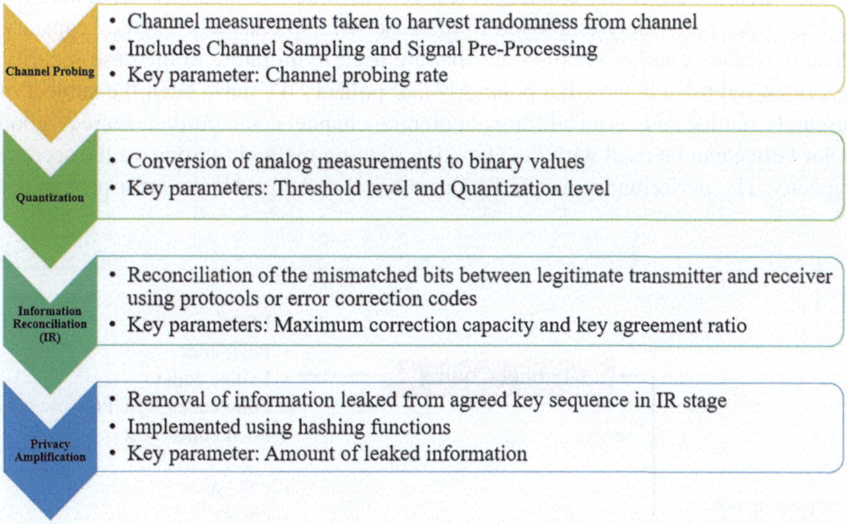

Fig. 6 PLS secret key generation procedure

are more difficult to exploit even with the backing of enormous computational power, unlike the traditional cryptographic techniques in which an intruder can easily decrypt with the aid of sufficient computing power [9].

There are four steps in the key generation process: channel probing, quantization, Information Reconciliation (IR), and privacy amplification, as summarized in Fig. 6 [9, 10].

The cryptographic key generation is addressed by key-based PLS, which may rely on established and current symmetric encryption techniques to maintain confidentiality. This decoupling feature makes it possible to dynamically update the encryption key as needed and adjust the key strength. These factors make it more likely that key-based PLS will fit more readily into a practical overall security approach.

6.2 *Key-Less Approach*

Key-less PLS approaches, which do away with the requirement for conventional cryptographic keys, mark a paradigm change in wireless communication security. Its foundation is information theory, drawing on the concept of secrecy capacity provided by the seminal works of Shannon and Wyner. The different techniques used for implementing key-less PLS [11] are shown in Fig. 7.

Channel Coding Wireless system security is greatly enhanced by error control codes when the adversary channel is less reliable than the legitimate channel. The LDPC codes are widely employed to increase wiretap channels' secrecy capacity. Polar codes are another set of codes that are used in literature to increase security. Recently, hybrid schemes like polar-AN and polar-LDPC have been the subject of research. Lattice code is an additional potential channel code. Further, secrecy-good polar lattice can be used with the Gaussian shaping method to improve the secrecy capacity. The performance metric for the channel code used is the error probability

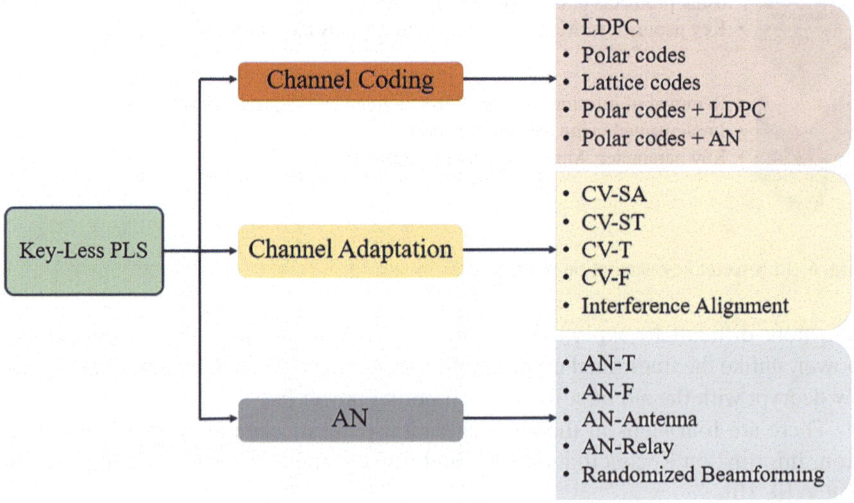

Fig. 7 Key-less PLS approaches

of codeword. A good channel coding has low error probability, leading to enhanced security.

Channel Adaptation When transmission parameters of the wireless system are dynamically adjusted and channel variations are exploited based on current channel conditions, performance is optimized and security is enhanced. This process is known as channel adaptation, or Channel Variation (CV). It is possible to use the independent fading that both the adversarial and legitimate channels undergo for this reason [12]. The various techniques under channel adaptation are

- Channel variation in spatial domain using antenna (CV-SA).
- Channel variation in spatial domain using relays (CV-SR).
- Channel variation in time domain (CV-T).
- Channel variation in frequency domain (CV-F).
- Interference alignment.

Artificial Noise (AN) The idea behind AN injection is to decrease the eavesdropper's capability to intercept and decode the information by adding artificial noise, like interference or jamming, into the transmitted signal. With prior knowledge of the noise pattern, the legitimate receiver can subtract it and retrieve the original message.

The authorized users have some freedom in such a setup where they can access null space. In a fading or nonfading environment, the perfect secrecy notion can be achieved with this method. The spatial, frequency, and temporal domains can all be used to introduce AN. The techniques under AN are

- AN in time domain (AN-T).
- AN in frequency domain (AN-F).
- AN in spatial domain using antenna (AN-Antenna).
- AN in spatial domain using relays (AN-Relay).
- Randomized beamforming in spatial domain.

6.3 Classification of WPLS Techniques

WPLS techniques are divided into three primary levels: Channel Level, Coding Level, and Signal Level, as shown in the Fig. 8 [7]. Each level includes various techniques and strategies that contribute to the overall security of the wireless communication system.

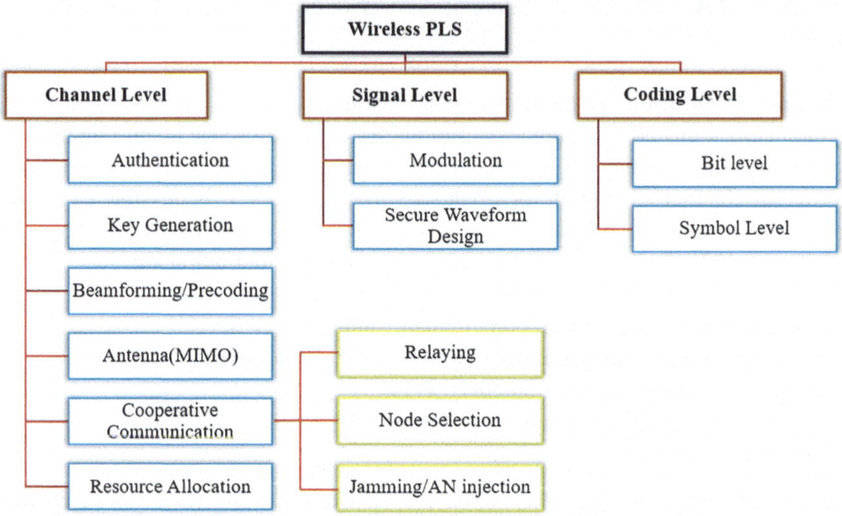

Fig. 8 Classification of wireless PLS

6.3.1 Channel Level

- **Resource Allocation:** This involves managing the resources (like power and bandwidth) to optimize the security of the communication channel. Proper allocation can minimize the risk of eavesdropping and other security threats.
- **Key Generation:** This is the process of creating cryptographic keys that are used to encrypt and decrypt data. Effective key generation methods enhance the security by ensuring that keys are difficult to predict or replicate.
- **Cooperative Communication:** This technique involves multiple nodes working together to improve communication security. By sharing information and resources, the overall resilience against attacks can be increased.
- **Antenna Selection:** Choosing the optimal set of antennas for transmission and reception can improve the security of the communication link. This technique can help in minimizing the interception of the signal by unauthorized parties.
- **Authentication:** Ensuring that the communicating parties are genuine and authorized. Authentication mechanisms prevent unauthorized access and ensure that only legitimate users can access the communication channel.
- **Relaying:** Utilizing intermediate nodes to relay the signal can enhance security by making it harder for eavesdroppers to intercept the communication. This technique can also involve secure relaying methods to further protect the transmitted data.
- **Node Selection:** Choosing specific nodes for communication based on their security properties. By selecting the most secure nodes, the overall security of the communication network can be improved.

- **Jamming/AN Injection:** Employing techniques to disrupt the communication of unauthorized users. Jamming can block unauthorized signals, while AN injection can confuse eavesdroppers and enhance the security of the transmitted data.

6.3.2 Signal Level

- **Modulation:** The process of varying a signal to encode information. Secure modulation techniques can help in making the signal less susceptible to interception and unauthorized decoding.
- **Secure Waveform Design:** Designing waveforms that are inherently secure and resistant to attacks. Secure waveform design can make it more difficult for adversaries to detect or jam the communication signal.

6.3.3 Coding Level

- **Bit Level:** Security techniques applied at the bit level include error correction codes and encryption at the bit level. These techniques ensure that even if some bits are intercepted or altered, the overall message remains secure and intact.
- **Symbol Level:** At this level, symbols (groups of bits) are encoded in a secure manner to protect against interception and unauthorized decoding. Symbol-level security techniques can provide an additional layer of protection beyond the bit level.

7 Performance Metrics in PLS

Assessing and measuring the secrecy performance of a security system or technique using an appropriate metric is among the most important actions that must be taken after designing it. PLS approaches are evaluated using a set of performance measures that are well-defined. These measures aid in comprehending the degree of security offered by PLS techniques, their effect on communication efficiency, and their applicability in actual use cases [8].

As illustrated in Fig. 9, the secrecy metrics employed in the literature can be divided into two main types: SINR-based metric, which is linked to key-less PLS techniques, and complexity-based metric, which is linked to key-based PLS methods.

Secrecy Rate/Capacity

The maximum rate at which secure wireless connection may be established while guaranteeing that an eavesdropper cannot obtain any valuable information is known as secrecy capacity, C_S. A more secure communication channel is indicated by a higher secrecy capacity. The mathematical equation of secrecy capacity is

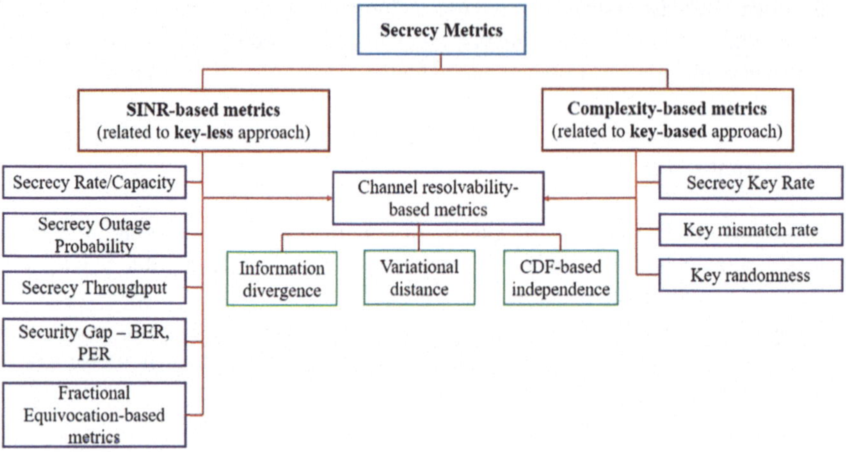

Fig. 9 PLS secrecy metrics

$$C_S = \max\left[0, C_B - C_E\right]$$

where C_B and C_E are the channel capacities of legitimate user (Bob) and eavesdropper (Eve), respectively.

Secrecy Outage Probability

The probability that the instantaneous secrecy capacity, C_S, drops below a targeted secrecy rate, R_S, and the intended level of security is not achieved is known as the secrecy outage probability, P_{out}. The measure of reliability of achieving secure communication is given by this metric. A more reliable and secure communication channel is indicated by a lower P_{out}. Secrecy outage probability is expressed as

$$P_{\text{out}} = P\left(C_S < R_S\right)$$

Bit Error Rate (BER)

The rate at which errors in transmitted bits occur and compromise the communication link's reliability is known as bit error rate, or BER. BER is measured at the eavesdropper as well as the legitimate user in the context of PLS. The main objective is to minimize BER for the legitimate user while maximizing it for the eavesdropper

$$BER = \frac{\text{Number of erroneous bits}}{\text{Total number of transmitted bits}}$$

Traditional secrecy outage probability metric has three major drawbacks, despite its use in assessing and characterizing the wireless channels' security performance. First, when outage secrecy occurs, it is unable to characterize the amount of information that leaks to the eavesdroppers quantitatively. Second, it is unable to offer any information regarding the eavesdropper's proficiency in deciphering the confidential information. Third, it is not compatible with the various applications' and services' Quality of Service (QoS) standards.

Inspired by these findings, three new metrics were proposed in [13] based on the distribution of partial secrecy, or fractional equivocation, provided by $\left(\Delta = \frac{H(M/R_E)}{H(M)}\right)$, which may be determined from channel gain distributions. These metrics consist of average fractional equivocation, average information leakage rate, and generalized secrecy outage probability. The complexity-based metric is mostly utilized in key-based systems since, with enough time and processing capacity, an eavesdropper may subsequently be able to figure out the key using an exhaustive search procedure, often known as a brute-force attack.

Secrecy Key Rate

The rate at which secure keys can be produced and shared between authorized parties via a wireless channel without being picked up by an eavesdropper is known as the secrecy key rate:

$$\text{Secrey Key Rate} = \frac{\text{Number of secret bits}}{\text{Time duration}}$$

The factors affecting secrecy key rate are channel conditions, SNR, and Channel State Information (CSI) estimation accuracy. Higher secrecy key rates are indicative of a more effective system that can produce secure keys more quickly, improving the communication system's overall security and responsiveness.

Key Mismatch Rate

In a PLS system, the key mismatch rate measures the frequency of differences between the keys generated by authorized parties. Effective secure communication requires a lower key mismatch rate:

$$\text{Key Mismatch Rate} = \frac{\text{Number of mismatched key bits}}{\text{Total number of generated key bits}}$$

Key Randomness

The unpredictability and entropy of the keys generated in a PLS system are measured by key randomness. A high degree of unpredictability makes the keys hard for an adversary to figure out or duplicate. Furthermore, the secrecy of key-based and key-less approaches can be measured using channel resolvability-based metrics, such as variational distance, information divergence, and CDF-based independence between the transmitted data and its observation at Eve.

Information Divergence

Information divergence quantifies the difference between two probability distributions and is often referred to as Kullback–Leibler (KL) divergence. In the context of PLS, it measures the degree to which the eavesdropper's observed signal distribution deviates from the intended distribution. The mathematical representation of KL divergence is

$$D_{KL}(P \| Q) = \sum_i P(i) \log \frac{P(i)}{Q(i)}$$

where P and Q denote the probability distribution of legitimate signal and eavesdropper's observed signal. High KL divergence indicates that the eavesdropper's observed signal distribution is significantly different from the legitimate distribution, implying better security.

Variational Distance

The maximum difference between the probabilities assigned by two distributions is measured by variational distance. It offers a precise indicator of the extent to which the legitimate signal can be distinguished from the eavesdropper's observations.

$$D_{var}(P,Q) = \frac{1}{2} \sum_i |P(i) - Q(i)|$$

Low variational distance indicates that the eavesdropper's observations closely resemble the legitimate signal and hence weaker security.

CDF-Based Independence

The CDF-based independence metric compares the cumulative distribution functions (or CDFs) of the two signals to determine how independent the eavesdropper's observed signal is from the legitimate signal:

$$D_{CDF} = \max_x |F_P(x) - F_Q(x)|$$

where F_P and F_Q are the CDFs of the legitimate signal and the eavesdropper's observed signal, respectively. High CDF-based independence represents that the CDFs of the eavesdropper's and legitimate signals are significantly different, suggesting higher security. By understanding and optimizing these metrics, researchers and engineers can develop more secure and effective PLS solutions.

8 PLS in Emerging Technologies

PLS's ability to improve wireless communication networks' overall security has drawn a lot of interest in the context of emerging technologies. PLS, in contrast to conventional cryptography techniques, uses the physical medium's intrinsic properties to protect data transmission. This section explores the applications of PLS in the emerging technologies [8] (Fig. 10).

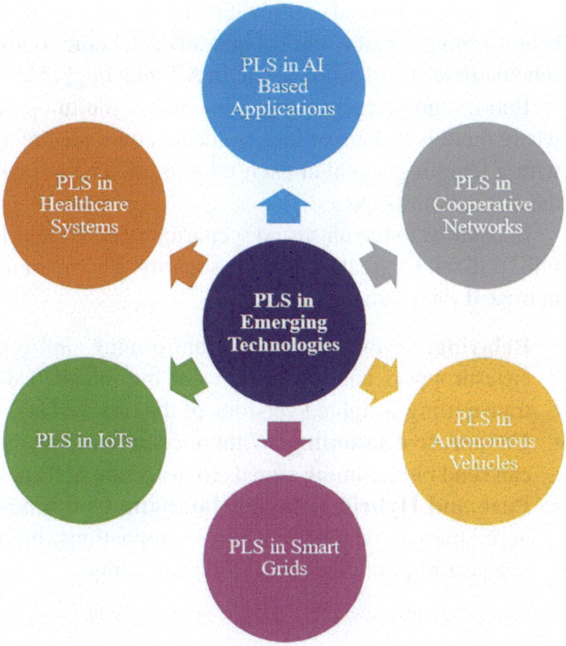

Fig. 10 Application of PLS in emerging technologies

8.1 *PLS in Cooperative Networks*

A cooperative communication network is one in which two or more relay nodes facilitate information flow between a source and a destination. The two protocols: the Decode-and-Forward (DF) and Amplify-and-Forward (AF) protocols are used by the relay nodes to process the signal that is received from the source. This network has proven to be an effective way to increase the capacity and coverage of single-antenna systems.

Relay nodes (untrusted and trusted) can be employed in wireless PLS to exploit the physical layer characteristics of wireless channels to enable secure transmissions from sources to destinations when eavesdroppers are present [14].

In cooperative networks, where the relay nodes are not entirely trusted to protect the confidentiality of the transmitted data, untrusted relays are used. In addition to helping to transmit information, these relays might attempt to decipher the source's private signal. Ad hoc networks, in which relay nodes are required for connectivity but are not authenticated, and networks inside government or financial institutions, where not all nodes have the same degree of security clearance, are two examples. Ensuring that these untrusted relays do not deduce the sensitive information being relayed, even if they aid in transmission, is the main challenge. Since DF needs the relay to decode the message based on its observation, it cannot be used in this case.

Through combined optimization of the relay beamforming matrix and the source beamforming vector, the cooperative scheme outperforms the noncooperative scheme in terms of secrecy rate in AF relaying [15].

End-to-end secrecy can be obtained in multihop communication circumstances where there is a chain of interconnected untrusted relays. The message is combined with a jamming signal at each relay in the chain, making it challenging to deduce the original message.

In contrast to the untrusted scenario, trusted relays have trusted nodes that can be fully utilized to greatly improve security. The main techniques to improve secrecy in trusted relays are as follows:

- **Relaying:** With distributed beamforming, information is directed away from eavesdroppers and steered toward the intended destination by multiple relays transmitting weighted versions of the received signal.
- **Cooperative Jamming:** Without compromising the intended user, trusted relays can send out jamming signals to lower the SNR at the eavesdropper.
- **Pure and Hybrid Relaying/Jamming Combinations:** Nodes can function as pure jammers or relays in pure combinations, but in hybrid combinations, they can perform both the roles at the same time.

8.2 *MIMO/Massive MIMO Systems*

Diversity techniques have been developed to boost system performance since the Shannon capacity restricts the capacity of single-antenna systems. These approaches use multiple replicas of transmission signals sent over many channels with different characteristics. Multiple antenna systems, also known as Multiple Input Multiple Output (MIMO) systems, provide very high spectrum efficiency and link dependability for wireless communications through space diversity. They also provide physical layer secrecy through novel and intriguing strategies. Secure beamforming, artificial noise generation, and relay-assisted secure transmission are some of the PLS strategies to improve security in MIMO systems [16].

Massive MIMO networks, in contrast to standard MIMO systems, use Base Stations (BSs) with a lot more antennas to transmit and receive simultaneously, providing faster speeds, higher service quality, and incredibly low latency to a large number of users. The performance of massive MIMO networks is, however, limited by certain adversarial attacks. When pilot sequences are reused across cells, interference results, making Pilot Contamination (PC) one of the most harmful attacks. The entire system performance is negatively impacted by this interference, which also reduces the effectiveness of channel estimation. To address this PC, pilot decontamination techniques can be used. The decontamination techniques include intelligent pilot reuse and allocation, blind channel estimation that rely on statistical properties of channel instead of pilots to estimate the CSI, and adaptive pilot assignment based on user locations and existing network conditions to reduce interference. Also, advanced signal processing techniques and distributed algorithms can be used to address the PC issue.

Further, cross-layer security is recognized as the future research area, wherein a strong, multilayered security framework for MIMO and massive MIMO systems can be achieved by integrating PLS techniques with cryptographic methods at higher layers.

8.3 Heterogeneous Networks (HetNets)

The existing Homogeneous Network (HomNet) can no longer satisfy the demands of the growing user base by offering enough services. Therefore, HetNets are regarded as the optimal solution for satisfying user requirements. HetNets, or heterogeneous networks, are formed when multiple network types—such as Wi-Fi, radio access networks, macro sites, and in-building wireless systems—are integrated to create a cohesive and effective network infrastructure. The diversified and dense distribution of these networks and varying coverage of different cells present PLS with distinct opportunities and difficulties (Fig. 11).

The key strategies developed to improve the security of HetNets are as follows:

- Secure beamforming includes adaptive beamforming which minimizes signal leakage to eavesdroppers while directing transmission energy toward authorized users using antenna arrays and cooperative beamforming where multiple cells or nodes work together to construct beamforming patterns that increase the signal strength at the intended recipient and decrease it at possible eavesdroppers.
- Artificial noise injection encompasses two types of noise injection: cooperative jamming, in which nodes from different cells jointly emit jamming signals to prevent eavesdroppers from intercepting transmitted data, and cell-specific noise, in which BS injects artificial noise specifically designed to confuse the eavesdroppers.
- The concept of resource allocation encompasses two main methods: dynamic resource allocation, which adjusts resource allocation strategies based on actual network conditions to maximize security, which includes modifying power levels and frequency channels, and interference management, which arranges resource allocation among various cells to control interference and make security improvements.
- Relay-assisted PLS using trusted and untrusted relays can also be used to improve secrecy in HetNets.

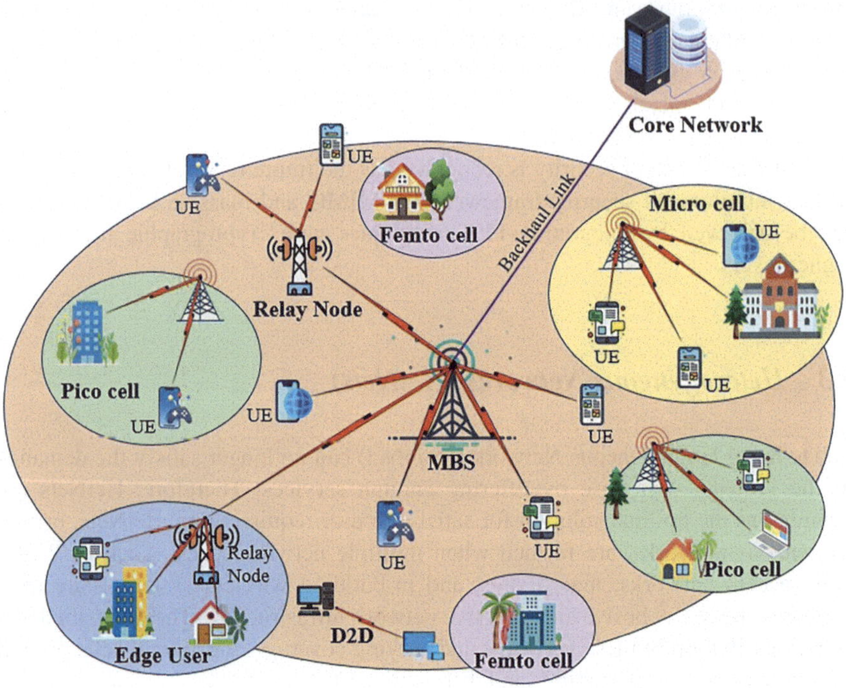

Fig. 11 Multi-tier 5G HetNet

8.4 PLS in Smart Grids

Similar to other wireless technologies, there are certain security vulnerabilities with the smart grid systems. Diverse communication technologies such as Wireless Sensor Networks (WSNs), Power Line Communications (PLC), and wireless mesh networks utilized by smart grids pose major security risks at the physical layer. Due to the broadcast nature of wireless channel and the critical infrastructure involved, smart grids are also susceptible to spoofing, jamming, and eavesdropping attacks. Therefore, by taking advantage of these vulnerabilities, attackers can interfere with grid operations, steal confidential data, or introduce erroneous data.

Many PLS solutions were proposed to enhance the security in smart grid. Some of them include adaptive modulation and coding based on CSI to improve the resilience against eavesdropping, a novel method for detecting malicious nodes based on the channel response of the node to identify fake smart meters, new signaling that creates a secret pattern which is only known by the sender and the recipient, a smart tracking firewall scheme to effectively identify harmful incursions and thwart security attacks at the node. Also, secure beamforming, artificial noise injection, cooperative jamming, and trusted and untrusted relays can be used to provide security at the physical layer.

8.5 PLS in Autonomous Vehicles

Autonomous Vehicles (AVs), sometimes referred to as driverless automobiles or autonomous cars, are a major development in transportation technology that have the potential to completely revolutionize how products and people are moved. AVs function without the driver's direct input, and the driver is not required to continuously watch the road. The implementation of AVs has been made possible by 5G technology, which can deliver speed of up to 10 Gbps with an ultra-low latency of 1 ms (for regular cellular users). However, AVs are vulnerable to a range of security risks because they rely on wireless connection for safety, navigation, and control tasks [17, 18]. For navigation and decision-making, AVs rely on real-time data sharing. Safe and reliable operation depends on protecting the confidentiality and integrity of this data (Fig. 12).

Using PLS techniques like physical layer key generation, beamforming, cooperative jamming, artificial noise injection, secure modulation, and coding will either send very few radio signals to the eavesdroppers or the signals sent will be entirely different from the signals transmitted to the authorized receiver. PLS therefore makes sure that a distinct wireless channel is allocated to each sender and receiver. The host vehicle's secrecy capacity measurement establishes whether or not the host vehicle and the target vehicle securely communicate via the enclosed wireless

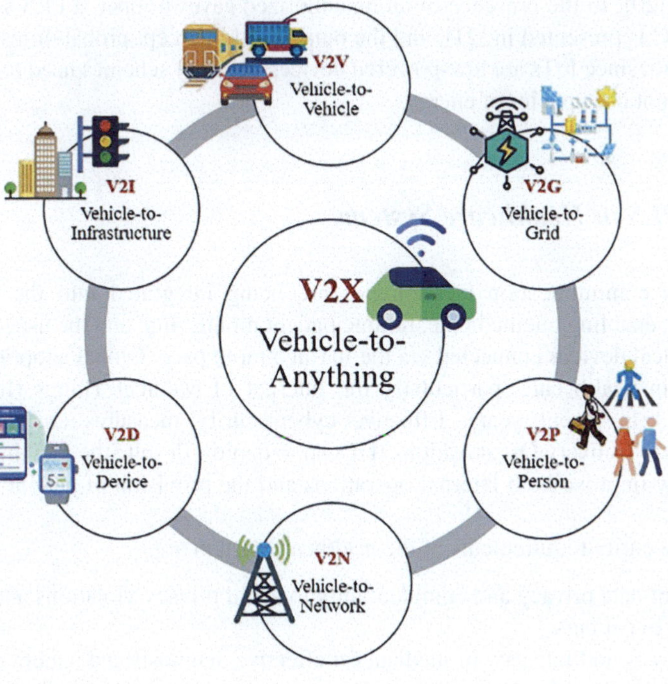

Fig. 12 V2X networks

channel. There are specific characteristics that impact secrecy capacity when employed only in V2V communications, unlike standard wireless communication channels.

8.6 PLS in Internet of Things (IoTs)

The Internet of Things (IoT), which enables ubiquitous connection of people, machines, and environments, is radically changing our way of life. Applications of IoTs are used in intelligent transportation, health care, smart homes and cities, and other fields. IoTs are low-power devices and have simple transmission strategies. The IoT devices frequently send sensitive data when interacting wirelessly. But the spread of these devices also brings with it serious security risks. To secure IoT communications, PLS provides creative solutions [19]. Further, one of the most promising and leading technologies for green IoT devices is Cognitive Ambient Backscatter Technology (C-AmBC).

Various novel PLS techniques are designed using the combination of AN injection, beamforming, secure encoding, and jamming to improve security. For example, a unique radio-frequency fingerprint can be created by extracting a slight random difference among transceivers to provide authentication [10], a novel space shift keying technique that uses AN injection and antenna selection to improve secrecy [20]. In the presence of an unauthorized eavesdropper, a PLS solution for C-AmBC is presented in [21], and the outage and intercept probabilities are evaluated. Also, since IoTs are low-powered devices, security schemes need to make sure they do not use up a lot of energy.

8.7 PLS in Healthcare Systems

Wireless communication technologies are being integrated into the healthcare industry, enabling telemedicine, remote patient monitoring, and the usage of different medical devices connected via the IoT at a rapid pace. Global adoption of technology in health care, particularly the Internet of Medical Things (IoMT), has increased in recent years. Effective cybersecurity measures are difficult for Healthcare Delivery Organizations (HDOs) to deploy, despite the fact that they significantly improve both internal operations and the provision of patient healthcare services.

The security requirements of the healthcare system are

- Patient data privacy and confidentiality to avoid privacy violations and potential harm to patients.
- Accuracy and integrity of medical for effective diagnosis and patient care.
- Compliance with stringent regulations to protect patient information.

The healthcare systems are affected by attacks like impersonation, jamming, and eavesdropping. PLS offers protecting wireless communications between medical devices and central monitoring systems using AN injection and jamming. By focusing signals on specific medical devices with beamforming, PLS reduces the risk of interception by unauthorized parties. Also, employing secure coding and modulation techniques like spread spectrum and Orthogonal Frequency Division Multiplexing (OFDM) in medical data transmission improves robustness against interference and eavesdropping.

The benefits of employing PLS techniques in healthcare systems include enhanced security, scalability, integration with existing technologies, real-time protection, and reduced latency. On the other hand, the complexity and cost of implementing PLS, highly dynamic healthcare environments, computational overhead, and lack of standard protocols for these systems pose challenges in designing efficient PLS algorithms.

8.8 PLS Using Artificial Intelligence (AI)

Artificial Intelligence (AI) is the idea and creation of computer systems that can do tasks that have traditionally required human intelligence, such as pattern recognition, speech recognition, and decision-making. It enables computers and machines to simulate human intelligence and problem-solving capabilities. Initially, AI was widely used for network traffic prediction and routing optimization in the upper layers of the OSI model. Later, with AI's remarkable capacity to learn complicated models, accomplish end-to-end optimization, and adjust to changing surroundings, it has emerged as a viable solution for communication systems' demanding performance needs on the wireless physical layer. AI encompasses a wide range of technologies, some of which are Machine Learning (ML), Deep Learning (DL), and Natural Language Processing (NLP).

ML is an area of AI that focuses on creating systems that can process and learn from data in order to make decisions. ML models are taught using data and algorithms to identify patterns and provide predictions or choices, as opposed to being expressly programmed to carry out a specific task. Figures 13 and 14 show some examples of how PLS schemes are implemented using ML-based neural and non-neural networks [7, 16].

ML-based algorithms can also be used in coding, modulation, and detection of channel estimation of wireless communication networks [33].

Channel Estimation

To achieve the desired Channel State Information (CSI) at the receiver, channel estimate is essential.

The complexity of channel estimate has increased due to new communication technologies because of large-scale matrix operations and fast CSI ageing in high-speed scenarios. Advances in machine learning and processing power have made AI

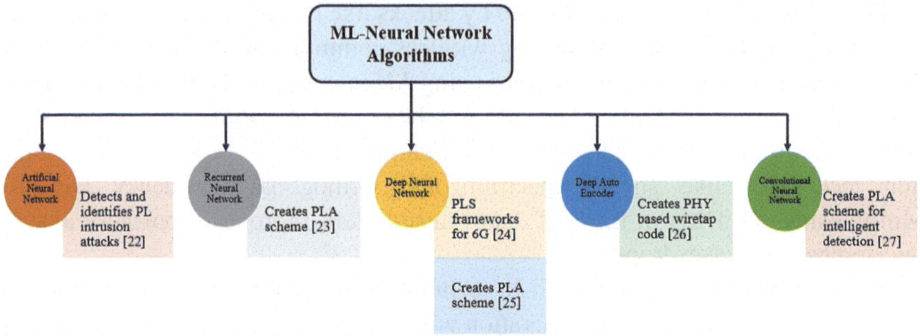

Fig. 13 ML—neural network algorithms

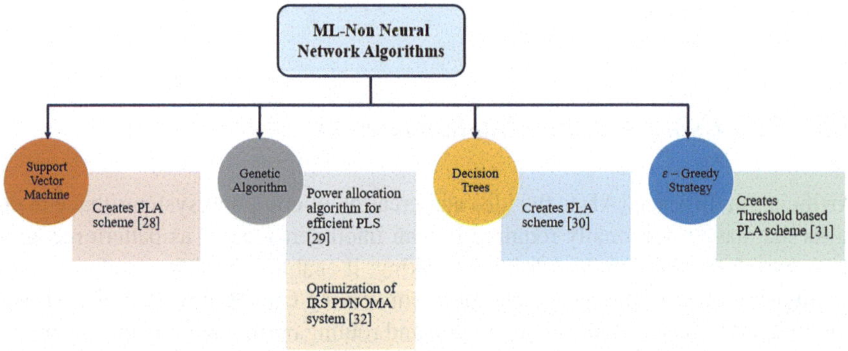

Fig. 14 ML—non-neural network algorithms

techniques feasible for improving the channel estimation performance [34]. AI-based channel estimation methods are used for synchronization, parameter estimation, channel prediction, and online training for estimation [35–39].

Through blind synchronization, AI can make up for the offset caused by the actual nonlinear channels. AI reduces the computational cost and synchronization errors associated with timing, carrier, and frame synchronization.

AI is able to efficiently capture the characteristics of the channel and enhance the accuracy of channel parameter estimation under a variety of channel situations, including doubly selective fading channels. AI is frequently used to estimate channel parameters for a variety of systems, such as OFDM, Filter Bank Multi-Carrier (FBMC), MIMO, and others. Furthermore, a notable area of research is the combined use of AI for signal recognition and parameter estimation.

AI, particularly Long Short-Term Memory (LSTM), can be used in rapidly changing channels to forecast future CSI based on the channel's recent conditions.

The channel estimation module is trained online using a small number of training samples using meta-learning and transfer learning techniques. This allows the module to quickly adapt to dynamic channels and resolves the issue of inconsistency between the training and actual models.

Coding

Coding which includes source coding and channel coding, generates information bits for transmission. AI improves source coding's efficacy by facilitating better feature extraction, which is essential in situations with complicated sources and constrained bandwidth [40]. Deep feedforward networks, Convolutional Neural Networks (CNNs), and LSTM networks are some of the approaches used in multimedium compression/recovery to reduce High-Efficiency Video Coding (HEVC) complexity and adjust to communication channels with limited bandwidth [41]. It has been demonstrated that AI techniques greatly minimize encoding delays and enhance source compression and quantization techniques [42].

AI provides a novel approach to channel coding that enables the creation of reliable channel decoders and end-to-end topologies [43]. As a result, decoding complexity is decreased and accuracy is increased, especially in intricate fading channels. Model-driven design, founded on conventional techniques such as Belief-Propagation (BP) algorithms, is widely used [44].

Autoencoders, often known as AEs, are used for end-to-end optimization of channel decoding. In the presence of AWGN and inter-symbol interference (ISI), a number of AE-based systems have proven to be able to handle non-white noise and enable arbitrary coding speeds [45, 46]. The NMSE performance of these techniques is better than that of conventional Bit-Interleaved Coded Modulation (BICM). Further, for Joint Source and Channel Coding (JSCC), deep AE networks are introduced to replace the conventional modules [47].

Modulation and Detection

Modulation involves bit-to-symbol mapping (achieved through constellation design) and symbol-to-waveform mapping (achieved through waveform design). Optimizing constellation points and decision regions and applying DNN-based approaches to improve spectral efficiency and adaptability are AI-enhanced transmitter design methods. For example, the constellation point distribution is optimized via a DL-based variational autoencoder in [48], and Turbo detection-aided multicarrier AE is designed for reliable soft-bit decisions in [49].

Similarly, a DNN structure replaces traditional serial hard decision receivers, improving BER in [50] under AI-enhanced detection. Also, signal detection of new modulation-based designs like OFDM-IM and GFDM-IM leverage AI to improve error performance and detection accuracy [51, 52]. AI-based enhanced techniques can also be used in relay transmission, multiple access, and MIMO.

9 Challenges and Future Directions of PLS

PLS is a developing field that aims to improve wireless data transmission security by taking advantage of the physical properties of communication networks. While PLS has many potential benefits, there are a number of issues that need to be resolved before its full potential may be realized. Furthermore, new communication paradigms and technological developments will probably influence PLS's future course. Although PLS provides novel methods for securing wireless communication, there are a number of issues that need to be addressed such as

- Channel knowledge and estimation (need of accurate CSI and CSI availability).
- Complexity and practical implementation (hardware constraints and computational overhead).
- Scalability to large-scale networks.
- Changing channels in mobile and dynamic environments.
- Interference management like friendly jamming.
- Trade-offs with other performance metrics (throughput and latency).
- Trustworthiness of cooperative nodes.

In the future, PLS's research and development will probably concentrate on establishing industry standards, optimizing cooperative methods, utilizing AI, combining PLS with upcoming technologies like 5G and beyond and quantum communications, and creating resource-efficient IoT solutions.

10 Conclusions

PLS has enormous potential for providing secure and safe wireless communication. Various security requirements of the OSI layers are explained. The foundation information theory for PLS, the approaches, and classification are discussed in the chapter. Further, the metrics of security measurement for PLS as well as the application of different emerging technologies are described. Finally, the challenges faced by PLS and its future research direction are outlined.

References

1. Kihero, A. B., Furqan, H. M., Sahin, M. M., & Arslan, H. (2024, June). 6G and beyond wireless channel characteristics for physical layer security: Opportunities and challenges. *IEEE Wireless Communications, 31*(3), 295–301. https://doi.org/10.1109/MWC.002.2300002
2. Shi, W., Jiang, X., Jinsong, H., Abdelgader, A. M. S., Teng, Y., Wang, Y., He, H., Dong, R., Shu, F., & Wang, J. (2022). Physical layer security techniques for data transmission for future wireless networks. *Security and Safety, 1*. https://doi.org/10.1051/sands/2022007
3. Shannon, C. E. (1949). Communication theory of secrecy systems. *The Bell System Technical Journal, 28*, 656–715.

4. Wyner, D. (1975). The wire-tap channel. *The Bell System Technical Journal, 54*, 1355–1387.
5. Lavanya, D. L., Ramaprabha, R., Thangapandian, B., & Gunaseelan, K. (2021). Novel privacy preserving authentication scheme based on physical layer signatures for mobile payments. *SN Computer Science, 2*, 119. https://doi.org/10.1007/s42979-021-00509-8
6. Sanenga, A., Mapunda, G. A., Jacob, T. M. L., Marata, L., Basutli, B., & Chuma, J. M. (2020). An overview of key technologies in physical layer security. *Entropy, 22*(11), 2020. https://doi.org/10.3390/e22111261
7. Kamboj, K., Jindal, P., & Verma, P. (2021). Machine learning-based physical layer security: Techniques, open challenges, and applications. Wireless Networks, 27, 5351–5383. https://doi.org/10.1007/s11276-021-02781-1
8. Hamamreh, J. M., & Arslan, H. (2017, November). Time-frequency characteristics and PAPR reduction of OTDM waveform for 5G and beyond. In Proceedings of the 10th international conference on electrical and electronics engineering (ELECO) (pp. 681–685).
9. Linh, D. V., Yem, V. V., Kien, T. V., & Thao, H. T. P. (2024). Key generation algorithms based on complex components of channel impulse response for massive MIMO wireless communication systems. IEEE Access, 12, 89947–89956. https://doi.org/10.1109/ACCESS.2024.3419052
10. Zhang, J., Rajendran, S., Sun, Z., Woods, R., & Hanzo, L. (2019, October). Physical layer security for the internet of things: Authentication and key generation. IEEE Wireless Communications, 26(5), 92–98. https://doi.org/10.1109/MWC.2019.1800455
11. Kumar, M. S., Ramanathan, R., & Jayakumar, M. (2022, November). Key less physical layer security for wireless networks: A survey. Engineering Science and Technology, 35. https://doi.org/10.1016/j.jestch.2022.101260
12. Rengaraj, R., Loganathan, L. D., & Konganathan, G. (2023, May). Secure upgraded spatial modulation for wireless transmission systems. Computer Networks, 226. https://doi.org/10.1016/j.comnet.2023.109680
13. Yusuf, M., & Arslan, H. (2016, November). Enhancing physical-layer security in wireless communications using signal space diversity. In Proceedings of the IEEE military communications conference (MILCOM) (pp. 1190–1194).
14. Rodriguez, L. J., Tran, N. H., Duong, T. Q., Le-Ngoc, T., Elkashlan, M., & Shetty, S. (2015, December). Physical layer security in wireless cooperative relay networks: State of the art and beyond. IEEE Communications Magazine, 53(12), 32–39. https://doi.org/10.1109/MCOM.2015.7355563
15. Zhang, Y., Zhao, S., Shen, Y., Jiang, X., & Shiratori, N. (2024). Enhancing the physical layer security of two-way relay systems with RIS and beamforming. IEEE Transactions on Information Forensics and Security, 19, 5696–5711. https://doi.org/10.1109/TIFS.2024.3402373
16. Irram, F., Ali, M., Naeem, M., & Mumtaz, S. (2022, October). Physical layer security for beyond 5G/6G networks: Emerging technologies and future directions. Journal of Network and Computer Applications, 206. https://doi.org/10.1016/j.jnca.2022.103431
17. Hakak, S., Gadekallu, T. R., Ramu, S. P., Parimala, M., Maddikunta, P. K. R., de Alwis, C., & Liyanage, M. (2022). Autonomous vehicles in 5G and beyond: A survey. Networking and Internet Architecture. https://doi.org/10.48550/arXiv.2207.10510
18. Ahn, N.-Y., & Lee, D. H. (2019). Physical layer security of autonomous driving: Secure vehicle-to-vehicle communication in a security cluster. Electrical Engineering and Systems Science, Signal Processing. https://doi.org/10.48550/arXiv.1912.06527
19. Huang, D., & Al-Hourani, A. (2024). Physical layer spoof detection and authentication for IoT devices using deep learning methods. IEEE Transactions on Machine Learning in Communications and Networking, 2, 841–854. https://doi.org/10.1109/TMLCN.2024.3417806
20. Sun, L., & Du, Q. (2017, December). Physical layer security with its applications in 5G networks: A review. *China Communications, 14*(12), 1–14. https://doi.org/10.1109/CC.2017.8246328
21. Li, X., et al. (2021, September). Physical layer security of cognitive ambient backscatter communications for green internet-of-things. *IEEE Transactions on Green Communications and Networking, 5*(3), 1066–1076. https://doi.org/10.1109/TGCN.2021.3062060

22. Natalino, C., Schiano, M., Di Giglio, A., Wosinska, L., & Furdek, M. (2019, August 15). Experimental study of machine-learning-based detection and identification of physical-layer attacks in optical networks. Journal of Lightwave Technology, 37(16), 4173–4182. https://doi.org/10.1109/JLT.2019.2923558
23. Germain, K. S., & Kragh, F., (2021). Mobile physical-layer authentication using channel state information and conditional recurrent neural networks. In 2021 IEEE 93rd vehicular technology conference (VTC2021-Spring) (pp. 1–6). https://doi.org/10.1109/VTC2021-Spring51267.2021.9448652.
24. Ara, I., & Kelley, B. (2024). Physical layer security for 6G: Toward achieving intelligent native security at layer-1. IEEE Access, 12, 82800–82824. https://doi.org/10.1109/ACCESS.2024.3413047
25. Alhoraibi, L., Alghazzawi, D., Alhebshi, R., & Rabie, O. B. J. (2023). Physical layer authentication in wireless networks-based machine learning approaches. Sensors, 4. https://doi.org/10.3390/s23041814
26. Besser, K. L., Lin, P. H., Janda, C. R., & Jorswieck, E. A. (2020). Wiretap code design by neural network autoencoders. IEEE Transactions on Information Forensics and Security, 15, 3374–3386. https://doi.org/10.1109/TIFS.2019.2945619
27. Chen, Y., et al. (2020). On physical-layer authentication via triple pool convolutional neural network. In 2020 IEEE globecom workshops (GC Wkshps, Taipei, Taiwan (pp. 1–6). https://doi.org/10.1109/GCWkshps50303.2020.9367391.
28. Yoon, J., Lee, Y., & Hwang, E. (2019). Machine learning-based physical layer authentication using neighborhood component analysis in MIMO wireless communications. In 2019 International conference on information and communication technology convergence (ICTC) (pp. 63–65). https://doi.org/10.1109/ICTC46691.2019.8939862.
29. Okati, N., Mosavi, M. R., & Behroozi, H. (2016). A genetic approach in relay-jammer selection and power allocation for physical layer security. In 2016 8th International symposium on telecommunications (IST) (pp. 374–379). https://doi.org/10.1109/ISTEL.2016.7881844.
30. Kuruvatti, N. P., Mallikarjun, S. B., Kusumapani, S. C., Djuitcheu, H., & Schotten, H. D. (2023). Physical layer authentication in private campus networks based on machine learning. In 2023 IEEE 97th vehicular technology conference (VTC2023-Spring) (pp. 1–6). https://doi.org/10.1109/VTC2023-Spring57618.2023.10201068.
31. Pan, F., et al. (2017). Physical layer authentication based on channel information and machine learning. In 2017 IEEE conference on communications and network security (CNS) (pp. 364–365). https://doi.org/10.1109/CNS.2017.8228660.
32. Kieu-Xuan, T., Nguyen-Thi, H., & Le-Thi, A. (2024). Multi-objective optimization-based GA in PLS of IRS-assisted PDNOMA communication. IEEE Access, 12, 87361–87383. https://doi.org/10.1109/ACCESS.2024.3417619
33. Pagidimalla, P., Venketbabu, T., Kanimozhi, J. K., Renuka, S., Anuradha, N., & Surenderkumar, S. (2023). Data security analysis of PHY layer using artificial intelligence and machine learning. In 2023 9th international conference on advanced computing and communication systems (ICACCS) (pp. 2507–2515). https://doi.org/10.1109/ICACCS57279.2023.10112749.
34. Ma, K., Gao, Z., Wang, J., & Cheng, L. (2023). Physical layer security design for FDD IM-OTFS transmissions based on secure mapping. IEEE Access, 11, 98293–98304. https://doi.org/10.1109/ACCESS.2023.3312580
35. Ye, H., Li, G. Y., & Juang, B. H. (2018, February). Power of deep learning for channel estimation and signal detection in OFDM systems. IEEE Wireless Communications Letters, 7(1), 114–117. https://doi.org/10.1109/LWC.2017.2757490
36. Dai, L., & Wei, X. (2022, July). Distributed machine learning based downlink channel estimation for RIS assisted wireless communications. IEEE Transactions on Communications, 70(7), 4900–4909. https://doi.org/10.1109/TCOMM.2022.3175175
37. He, H., Wang, R., Jin, W., Jin, S., Wen, C. K., & Li, G. Y. (2023, March). Beamspace channel estimation for wideband millimeter-wave MIMO: A model-driven unsupervised learning approach. IEEE Transactions on Wireless Communications, 22(3), 1808–1822. https://doi.org/10.1109/TWC.2022.3206773

38. Banerjee, B., Elliott, R. C., Krzymień, W. A., & Farmanbar, H. (2023, January). Downlink channel estimation for FDD massive MIMO using conditional generative adversarial networks. IEEE Transactions on Wireless Communications, 22(1), 122–137. https://doi.org/10.1109/TWC.2022.3191605
39. Weththasinghe, K., Jayawickrama, B., & He, Y. (2024, April). Machine learning-based channel estimation for 5G new radio. IEEE Wireless Communications Letters, 13(4), 1133–1137. https://doi.org/10.1109/LWC.2024.3362963
40. Van Huynh, N., et al. (2024, June). Generative AI for physical layer communications: A survey. IEEE Transactions on Cognitive Communications and Networking, 10(3), 706–728. https://doi.org/10.1109/TCCN.2024.3384500
41. Xu, M., Li, T., Wang, Z., Deng, X., Yang, R., & Guan, Z. (2018, October). Reducing complexity of HEVC: A deep learning approach. IEEE Transactions on Image Processing, 27(10), 5044–5059. https://doi.org/10.1109/TIP.2018.2847035
42. Manouchehri, S., Haghighat, J., Eslami, M., & Hamouda, W. (2022, June). A delay-efficient deep learning approach for lossless turbo source coding. IEEE Transactions on Vehicular Technology, 71(6), 6704–6709. https://doi.org/10.1109/TVT.2022.3155545
43. Ye, N., Miao, S., Pan, J., Ouyang, Q., Li, X., & Hou, X. (2024, June). Artificial intelligence for wireless physical-layer technologies (AI4PHY): A comprehensive survey. IEEE Transactions on Cognitive Communications and Networking, 10(3), 729–755. https://doi.org/10.1109/TCCN.2024.3382973
44. Liang, F., Shen, C., & Wu, F. (2018, February). An iterative BP-CNN architecture for channel decoding. IEEE Journal of Selected Topics in Signal Processing, 12(1), 144–159. https://doi.org/10.1109/JSTSP.2018.2794062
45. Zhang, Y., Wu, H., & Coates, M. (2022, February). On the design of channel coding autoencoders with arbitrary rates for ISI channels. IEEE Wireless Communications Letters, 11(2), 426–430. https://doi.org/10.1109/LWC.2021.3131848
46. Lu, Q., Xu, R., Liu, C., Li, S., & Bai, B. (2023). An unequal error protection-based coded transmission for federated learning. In 2023 IEEE/CIC International conference on communications in China (ICCC Workshops) (pp. 1–5). https://doi.org/10.1109/ICCCWorkshops57813.2023.10233736.
47. Chen, D., Lei, M., Zhao, M. -M., Liu, A., & Sheng, S. (2023). Deep learning based coded over-the-air computation for personalized federated learning. In 2023 IEEE 98th vehicular technology conference (VTC2023-Fall) (pp. 1–5). https://doi.org/10.1109/VTC2023-Fall60731.2023.10333645.
48. Ma, Z., Wu, W., Jian, M., Gao, F., & Shen, X. (2022, March). Joint constellation design and multiuser detection for grant-free NOMA. IEEE Transactions on Wireless Communications, 21(3), 1973–1988. https://doi.org/10.1109/TWC.2021.3108666
49. Xu, C., et al. (2022, June). Turbo detection aided autoencoder for multicarrier wireless systems: Integrating deep learning into channel coded systems. IEEE Transactions on Cognitive Communications and Networking, 8(2), 600–614. https://doi.org/10.1109/TCCN.2022.3168725
50. Zheng, S., Chen, S., & Yang, X. (2021, March). DeepReceiver: A deep learning-based intelligent receiver for wireless communications in the physical layer. IEEE Transactions on Cognitive Communications and Networking, 7(1), 5–20. https://doi.org/10.1109/TCCN.2020.3018736
51. Zhu, Y., Wang, B., Li, J., Zhang, Y., & Xie, F. (2022, November). Y-shaped net-based signal detection for OFDM-IM systems. IEEE Communications Letters, 26(11), 2661–2664. https://doi.org/10.1109/LCOMM.2022.3201113
52. Turhan, M., Öztürk, E., & Çırpan, H. A. (2019). Deep convolutional learning-aided detector *for generalized frequency division multiplexing with index modulation*. In 2019 IEEE 30th annual international symposium on personal, indoor and mobile radio communications (PIMRC) (pp. 1–6). https://doi.org/10.1109/PIMRC.2019.8904193.

Analysis of Cyberattacks on IOT Devices in Smart Homes

Jian Wu, Manoj Kumar, and May El Barachi

1 Definitions and Assumptions

According to the survey on the cybersecurity of smart grids, the distinct attacks can be considered as follows [14]:

- Denial of Service (DoS): Overloading the network machine by dispatching an excessive volume of information, instigating a system collapse.
- Spoofing: Pretending to be a legitimate node to compromise the system.
- Man-in-the-Middle Attacks: Preventing or altering data during its transit across the network.
- Brute Force Attacks: Cracking user identifications and associated passwords.
- Traffic Analysis Attack: Controlling the hosts and the devices that are connected to the network.
- Social Engineering Attacks: Strategically influencing users to disclose confidential data.

2 Introduction

With the development of science and technology, more and more advanced techniques were brought to human's lives for ease of living. The Internet of Things (IoT) and smart home play a significant role today. People's daily lifestyle has changed considerably by adopting them. Recently, IoT devices have been implemented

J. Wu · M. Kumar · M. El Barachi (✉)
School of Computer Science, University of Wollongong in Dubai, Dubai Knowledge Park, Dubai, UAE
e-mail: jw807@uowmail.edu.au; manojkumar@uowdubai.ac.ae; maielbarachi@uowdubai.ac.ae

© The Author(s), under exclusive license to Springer Nature Switzerland AG 2025
A. Mishra et al. (eds.), *Hardware Security: Challenges and Solutions*,
https://doi.org/10.1007/978-3-031-81213-2_9

widely in several areas, including but not limited to, domestic automation, urban informatics, entertainment, education, and so on [13]. The mechanism of smart home is the integration of a broad spectrum of devices, regardless of their scale and size, which establish connections to their respective networks [9].

At the same time, due to the excessive usage of IoT devices and smart homes, security vulnerability concerns have been raised by the public. It provides an opportunity to unethical hackers to hack and control these devices and systems remotely as all of them relate to network. According to the Kaspersky IoT cyberattack report, during the 6-month span from January to June 2021, a total of 1.51 billion breaches were reported concerning Internet-of-Things-centric devices, marking an increase of 639 million when compared with the equivalent time frame in 2020. Since many devices were connected, the IoT system might be unsafe because of the purposeful attack from hackers; what's more, some random failure would be caused like device down, communication interrupted, etc., and the interaction is unforeseen among these devices [18, 25]. According to the points of view of Lin and Bergmann [16], there are two major challenges of security threats: the unwariness of security knowledge and recognizing the amalgamation and harmonization of diverse technological modalities. Several researchers have already shared insights and put in their studies by addressing security risk. It provides the inspection about security bug and potential cybersecurity attacks. Given that the network-based smart home could have a bad impact on society and might be used for some illegal activities, multiple authentication methods like password, SMS, and biometric were introduced to deal with the issue. Authentication of users, which serves as a crucial gateway to comprehensive smart home systems, is often subconsciously overlooked, regrettably undermining the effectiveness of this verification method. If it was compromised, then the person with bad intention can easily access IoT devices to steal personal information or cause financial, physical, or mental loss. Additionally, the firewall's capability to filter and thwart malware intrusions is significant [24]. As is well-known, firewalls play a crucial role in preventing unauthorized external access to networks. Nonetheless, security risks may arise due to the disparity between the simple rules established by network administrators and the absence of standards governing the behavior of malware created by algorithms.

Based on the above analysis of cyberattacks on IoT and smart homes, a couple of methods could be implemented to mitigate those vulnerabilities. For instance, from the device's perspective, in a research paper [26], an agent-based method of fuzzy logic was utilized to detect the network sensors of smart homes. And another device-centric, domain-independent approach was proposed by Chen and Helal [8] to fix the issue that relates to security in IoT systems. Also, a safety assurance framework which was enabled by security was proposed to cover the three aspects: hardware, communication, and system model in IoT applications [13]. Apart from this, there are some authentication models that can be used to improve the security of smart homes. The shared secret authentication which is integrated with cryptograph can protect the data or information transfer between the sender and receiver. The other one is public key/digital certificate authentication which uses two different keys, public key and personal key. The former key is accessible to anyone, while the personal key is encrypted to accept the information [1].

This study primarily emphasizes a comprehensive analysis of the IoT and smart home sector, with a focus on addressing various cybersecurity attack methods. Additionally, it aims to explore the corresponding solutions and methods to effectively mitigate associated risks. However, there is a limitation due to the complexity of various IoT devices and different ways how the different equipment were implemented in smart homes. In other words, the solution may vary up to how consumers use the devices.

1. Cyberattacks on the IOT devices that are used in smart homes are increasing exponentially. How can these attacks be analyzed and mitigated?
2. With the advancement in technology, IOT devices used in smart homes, if not updated timely can become vulnerable to bad actors. What can we do to avoid such circumstances/threats from bad actors?

3 Problem Statement

3.1: Cyberattacks on the IOT devices that are used in smart homes are increasing exponentially. How can these attacks be analyzed and mitigated?

3.2: With the advancement in technology, IOT devices used in smart homes, if not updated timely, can become vulnerable to bad actors. What can we do to avoid such circumstances/threat from bad actors?

4 Research Questions

- Are smart home IoT devices susceptible to specific types of attacks?
- How can the attacks on IOT devices used for smart homes be mitigated?

5 Feasibility and Research Objectives

This study is feasible as many cities around the world already implemented IoT devices either in smart homes or smart cities. Especially for major cities like Shanghai, Beijing in China, the residences use this advanced technology to make life more effective. This study can help avoiding any cybersecurity risk to protect human and society safely.

The more IoT devices are implemented in smart homes, the more vulnerabilities or loopholes could be exposed to unethical hackers. It is needed to analyze these types of cybersecurity attacks to mitigate the potential risks, which might make these devices vulnerable, and this study is to mitigate these potential risks and avoid the users of IoT devices from different types of losses that may happen due to these cyberattacks or unawareness of users of these cyber threats.

6 Literature Review

Nowadays, human life is getting more and more convenient as well as complex with the advancement of IoT devices which are being implemented in smart homes and creating an environment that consists of many electrical, electronic, and computing devices that are hardware, user interface, web application, mobile application, network, and cloud as well as an API, as suggested by Akhilesh et al. [3]. This environment created for the convenience of human beings undoubtedly is providing much convenience in life, but along with that it produces many complexities such as these devices being controlled by someone that can disrupt the convenience of life. For example, the distance between the homeowner and the door can be preset and monitored by a camera, and once the value reaches a specific number, a command is initiated on smart devices like air conditioner and smart TV that can be turned on in advance to create a comfortable living environment for the family member. Meanwhile, the light and Bluetooth audio could be switched off after a certain time like 10 or 11 pm or at the suggested rest time. Moreover, all the IoT devices and the related system will be managed and controlled by a central hub; as a result, the different devices can work efficiently to benefit residents' life. However, [1] it is pointed that the increasing prevalence of smart homes has elevated cybersecurity concerns to a pivotal standpoint. Due to the nature of smart homes and IoT devices which are implemented through a computer network, compared to the traditional applications which were used in major operating systems, such as Windows, IOS, or Mac OS, the security of IoT encounters more challenge to protect the diverse devices on network from cyberattackers. Chifor [9] also emphasized the importance of accessing, monitoring, and controlling smart products. In terms of the smart home ecosystem, it is important that communication among various components should be interacted in a trusted model. Besides, Kumar [20] mentioned that noneffective integration between smart gadgets would make the occupant exposed to the public, which might result in a wide spectrum of security issues like financial loss, data breach, health damage, and emotional trauma, even death. Akhilesh et al. [3] also provided with a solution for mitigating these risks from bad actors; they propose the automated penetration testing framework which provides with the automated penetration testing of devices used in this environment along with the common vulnerability scoring system. All the devices are not included in the framework, but random devices with high vulnerability scores are taken into consideration that had to undergo a test at a random period of time; another way of doing so is the use of PENTOS, a penetration testing tool, which comes built-in with Kali Linux, that performs testing such as brute force for password, airmon-ng for Wi-Fi or network connection attacks, and web scanning, but its drawback is that is does not look for vulnerabilities specifically but performs hacking test that shows the vulnerability on specific methodologies given by Visoottiviseth et al. [27].

Several security threats which might occur in smart homes have been analyzed by researchers. According to Darem, [1], the most challenging point of security is the lack of standards of IoT security. Many manufacturers show urgency in conducting and selling more devices to earn profit but show less focus on the security risk. Second, the software or hardware of devices becomes out of date which needs to be automatically upgraded. Also, if all the data which is generated and transferred from the device is stored on the data center and uploaded to cloud, then the end-to-end security cannot be guaranteed. Lastly, many users will prefer to use simple passwords or policies which will increase the chance for bad actors to hijack the system. To address these kinds of potential security risks and threats, Darem proposed an authentication mechanism which can be employed in the smart home ecosystem. Basically, there are different models when it comes to the mechanisms. Satapathy [22] mentioned a shared secret authentication which is called symmetric. During the authentication process, one side will post a question as the challenge and the other side needs to reply with the correct answer which is marked as response. The key concept is to protect the security risk from man-in-the-middle. Another mode which is named as public key/digital certificate authentication is considered as asymmetric [17]. It uses two separate keys, one is a public key which can be accessed by public, while the associate personal key is to verify the identification in cryptograph. Another model is the Hardware Security Module (HSM) which is described by Batalla [5]. HSM is an independent hardware module which can store and verify digital certificates. Another popular model is biometric authentication model [4] which is based on the user unique biometric features, such as fingerprint and face. Due to the advantages of physically unclonable functions (PUFs), this kind of authentication could be a feasible solution. Apart from this, different security features and protocols that are applied in smart homes have been compared by Darem [1]. Thereby, some countermeasures have been suggested as well. The first one is the principle of security which is called "Never trust; always verify". Intended cyberattackers are always trying to find a way to invade the IoT system to control the devices. It is needed to follow the always verify policy and test and monitor the related services to make sure the operation from the ecosystem is running normally. Another point is using Internet cable instead of WiFi. In case that some budget or component can only be connected to a wireless network, then the service set identifier (SSID) must be modified and using weak and simple passwords should be avoided. What's more, the authentication mechanism should be applied to smart homes and the software and hardware should be kept the latest to reduce the breach of security bugs.

Likewise, the safety issue in the IoT area is another significant concern which has risen in our daily life [13]. The fire which was caused by Samsung Galaxy Note mobile phone shows both physical damage and emotional hurt of users. Hence, the safeguarding of these smart devices should be taken into consideration while implementing a smart home system. Given this, a security-enabled safety framework was designed and presented by Kabir et al. [13]. The framework consists of two domains.

One is system design and analysis, and another is system operation. There are seven steps in the first domain, which starts from identifying the individual system that is adopted in a smart home to defining the safety goals. After that, specific operation states should be analyzed, and the corresponding actions in terms of safeguarding should be addressed. Then, any conflicts among independent systems should be addressed and actions performed accordingly to address contradictions. Finally, the model will be generated with the purpose of monitoring safety. Meanwhile, the second domain will focus on the detailed operation with three processes. First, the model is used to monitor all systems, and data and evidence are collected while needed. Second, the current safety is determined during the execution. Eventually related action is taken based on the disparate status which was detected before. The author implemented the framework based on the following cases (see Fig. 1): case 1 contains the sensors of a smoke detector and heat detector, respectively. And a leak detector was installed in the next room and forms case 2 with case 1.

Several statuses (ranging from S0 to S7) were defined during the implementation, and effective actions were also built to deal with safety issues. For instance, S0 defines that both smoke and heat detectors are working normally, and there is no need of action, and the alarm and sprinkler are just turned off. However, S3 describes that only the smoke detector is working and nothing else is detected; then, the action is informing the occupant regarding the failure of the heat detector. The framework

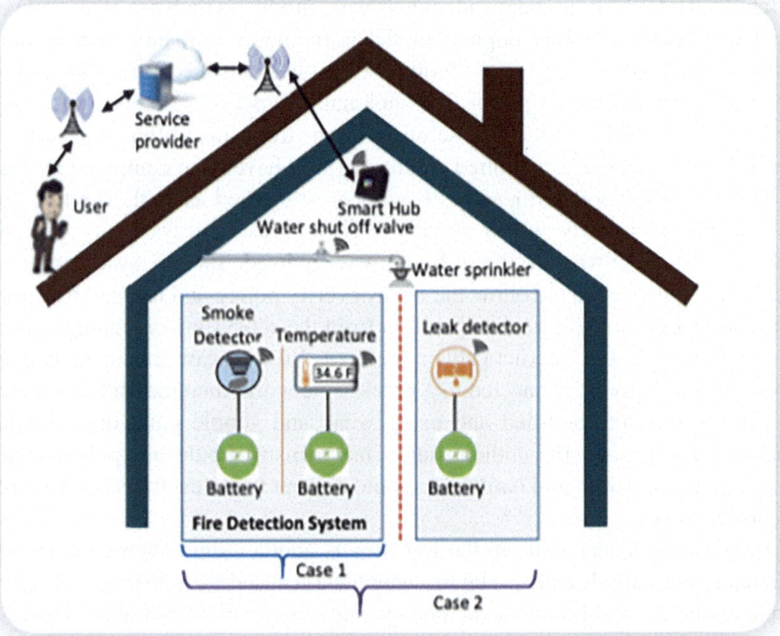

Fig. 1 IoT-based smart home environment [13].

has advantages, especially for monitoring all the smart components and sensors and taking workable action to mitigate the safety concerns. However, it might be complicated when it applies to big houses with plenty of devices.

Other types of cyber threats to smart homes are also investigated by researchers [10]. According to Stamm et al. [23], by using the phishing web page, the attackers can execute the Java application on the devices which will try to recognize the IP address of the router. In this situation, the router specifics will be addressed, and the product manufacturers will be attacked by hackers to get the default login credentials (username and password). If the user does not change it, then the devices can be compromised. WiFi de-authentication attack is a well-known breach which exists in protocol 802.11. The bug can be exploited to disconnect the wireless from the end device and steal the username and password from occupants who are unaware of the attack [7]. Near-Field Communication (NFC) is a useful technique which is widely used in access control in many areas, like the access cards for offices in business buildings, reception of hotels, and in public transportation. The vulnerability is that the NFC information can be monitored, and the data could be reproduced to open other devices which support the NFC card [19]. On the other hand, some measurements were proposed to fix all kinds of security vulnerabilities. Heartfield [10] found that anti-virus vendors already get into the smart home market in terms of security needs. Some vendors such as MacFee and Bit defender have already developed a platform to replace the current WiFi router and gateway. The concepts of security framework are similar regardless of the platform to analyze data which was sent to the vendor to protect the user's safety. Rahmati et al. [21] were aspired by an authentication mode on the new smart home platform. They suggested that access control should be based on risk instead of functionality. At the same time, they evaluated the new model based on the existing smart apps and found that the high-risk access was decreased by 60% with no impact on the service features or decision-making by the occupant.

7 Research Methodology

IOT devices' usage, as we know from the abovementioned literature, has increased exponentially; so, for that to deal, we have different domains of security in the IOT devices, namely hardware, network, and application part of IOT interconnections.

Given the deep-seated passion for networks and network security, coupled with foundational experience and knowledge in this domain, we are naturally inclined toward specializing in the network security of IoT devices and their integration within networks. The pathway toward achieving this will be as follows:

1. We will be using small experiments with few IOT devices interconnected.
2. We will be using network monitoring tools (such as NMAP and Wireshark) to detect the loopholes and open ports on the network as well as IOT devices.

3. A survey will be carried out to get information from general people of the social engineering threats to the IOT devices (this will be a general survey specific to the social engineering threats).
4. Then, we will look for the solutions toward the mitigation of these vulnerabilities.

To get started with, we are going to set up a smart home with a few IoT devices, like routers, smart speaker, watch, security camera, and sensors to evaluate the security of the network and identify the potential vulnerabilities. The control hub should be a smart speaker to control all smart home devices. Meanwhile, this will be connected on the network that was set up by the routers.

Before delving into the network monitor tools, let us have a brief review of NMAP and Wireshark. NMAP stands for Network Mapper; it is an open source as well as a free tool to scan ports and map networks to detect the vulnerabilities. According to Liao et al. [15], NMAP is a commonly utilized scanning tool that gathers data from the target host. The acquired information can be analyzed to aid in subsequent attacks. The tool provides a comprehensive set of features for network exploration and security auditing. It allows us to discover hosts, services, and open ports on a network. Wireshark is a network packet analysis software. The function of network packet analysis software is to intercept network packets and display the most detailed network packet information as possible. One of the most used tools in analyzing packets is Wireshark. Nevertheless, Wireshark is less effective in detecting the intrusion, still it can be used to intercept and detect network-encrypted traffic which will analyze the captured traffic [12].

The experimental setup can be established. The initial step involves configuring the various IoT devices on the network. Subsequently, NMAP is utilized to scan the network and identify the devices connected to it. Following this, the Command or Terminal tools, contingent upon the operating system in use, are opened to execute the following commands: "Nmap -sn <network IP range>." It is important to note that the network IP range corresponds to the range of the router's IP addresses. For Mac OS, version 13.3.1, this can be ascertained by following these steps: Navigate to System Settings, select WiFi, and identify the SSID currently used by the smart home. Adjacent to the SSID, locate the "Details" option, and click on it to access the TCP/IP tab. The IP range will then be displayed (see Fig. 2). NMAP will scan the specified IP range and present the active hosts within the network. Once the active hosts are identified, the subsequent phase involves in-depth analysis, including the exploration of open ports and running services. The NMAP command line for this purpose is as follows: "Nmap -p- <target IP address." The "target IP address" designates the specific IP address of the IoT device under investigation, which is crucial for obtaining information about open ports, services, and potential security vulnerabilities.

Meanwhile, Wireshark's analysis consists of two major processes. The first one is analyzing the network traffic. For doing this, (1) Launch Wireshark and select the network interface through which the IoT devices are connected. (2) Start capturing packets and perform activities on the IoT devices, such as turning on/off the smart

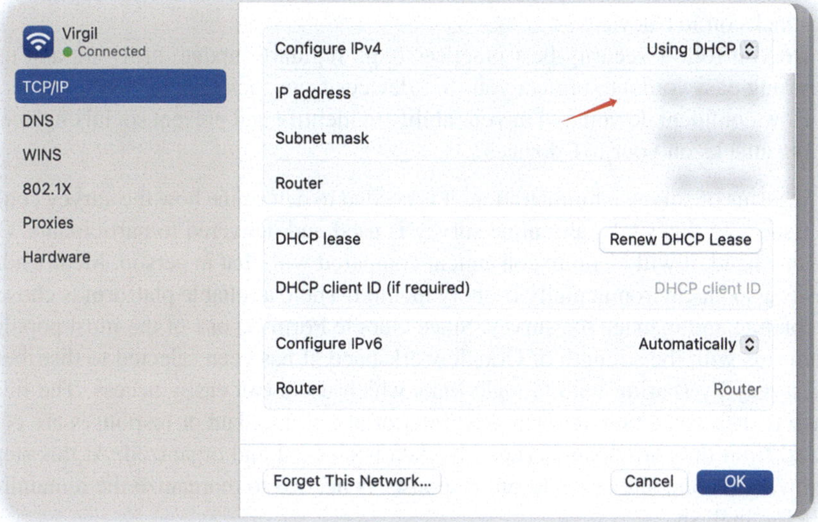

Fig. 2 IP address

bulbs or accessing the security camera feed. 3) Wireshark will capture and display the network traffic generated by these activities. The second step is to examine the packets in Wireshark. To achieve this, (1) Analyze the captured packets in Wireshark to address any suspicious or unauthorized activities. (2) After that, search for abnormal traffic patterns, unrecognized protocols, or any signs of malicious communication. Focus on the source and destination IP addresses, ports, and the contents of the packets. By combining the NMAP and Wireshark, we can gain insights about the network security of IoT devices in terms of vulnerability. NMAP assists in identifying the active hosts, open ports, and running services, while Wireshark allows to analyze and monitor the network traffic for any suspicious activities or security breaches.

Conducting a survey on social engineering threats specific to IoT devices can gather related information from the public regarding their awareness and experiences with social engineering attacks targeting IoT devices. This valuable information will help inform strategies and recommendations for enhancing security awareness and mitigating social engineering risks in IoT device usage. At the survey design stage, the author starts with defining the objectives of the survey, such as the specific outcomes which need to be achieved from the survey. Then the target audience is addressed, especially who will participate in the survey with different roles or positions in society (students, employees, and so on). Lastly, the detailed questions of the survey need to be addressed.

For example

- Are you familiar with social engineering attacks targeting IoT devices?
- Have you ever experienced a social engineering attack on your IoT devices?

- Are you aware of the potential risks and consequences of social engineering attacks on IoT devices?
- Do you follow security-best practices (e.g., regularly update firmware and use strong passwords) to protect your IoT devices from social engineering attacks?
- How confident do you feel in your ability to identify and prevent social engineering attacks on your IoT devices?

In terms of survey administration, it is needed to determine how the survey could be issued. In this study, an online survey is used and delivered to participants. On the one hand, it will be more convenient compared with that in person. Meanwhile, this is also an environmentally friendly method. Then, a reliable platform is chosen to conduct and manage the survey. Since Google Forms is one of the most popular platforms with the strength of Google workspace, it has been selected to distribute the survey by sharing with Google links which users can easily access. The final stage is data collection and analysis: First of all, all the data or responses are collected from the participants. Then, the data is cleaned and organized. At this step, removing incomplete or irrelevant responses is needed to reorganize the remaining data for analysis.

The next step is to analyze patterns, trends, and insights related to participants' awareness. Ultimately, we can extract the finding to understand the level of awareness, vulnerability, and potential areas of improvement regarding social engineering threats in IoT device usage.

Based on the analysis of small smart home network implementation with a few IoT devices, identifying network activities with IP addresses via using several tools, monitoring, and analyzing packets, the vulnerabilities specific to the IoT devices used in smart homes will be identified. On the other hand, summarizing the survey findings can provide an overview of the participants' perspectives on social engineering threats to IoT devices. Some common vulnerabilities and security challenges will be reported by users as well which can support us to conduct some solution toward the mitigation of these vulnerabilities.

8 Experimentation

In this chapter, a smart home is equipped with different IoT devices made in China. The brand Xiaomi, a Chinese technology company, has established a notable presence in the realm of smart home technology by offering a diverse selection of cost-effective and cutting-edge smart home products. Xiaomi's smart home ecosystem, often referred to as "MiJia," comprises several salient features:

- **Smart Home Devices:** Xiaomi's product portfolio encompasses a wide variety of intelligent devices, including but not limited to smart speakers, lighting systems, and surveillance cameras. These devices are meticulously designed to seamlessly integrate into a unified connected ecosystem.

- **Mi Home App:** The Mi Home application functions as the central control hub for Xiaomi's assortment of smart home devices. It empowers users to manage and monitor numerous devices from a singular interface, endowing them with a convenient and user-friendly experience.
- **Cost-Effectiveness:** Xiaomi has garnered a good reputation for providing top-tier smart home products at competitive price points. This affordability factor has substantially expanded the accessibility of smart home technology to a wider demographic of consumers.

The devices which were adopted are Xiaomi smart speaker (IR Control), bedside lamp, door and window sensor, Xiaomi motion sensor 2S, Mi flood detector, Xiaomi wireless switch and MI Home security camera. The detailed specification is listed in Table 1.

8.1 Installation

To pair all the smart devices with MIJIA, the application Mi Home is required to be installed on mobile devices. In this research, we selected iPhone 14 Pro max with IOS version 17.0.3 as the platform to install MI Home. To process, just go to App Store and search its name and then register an MI account with the mobile phone number. After that, we can start to configure the above IoT gadget with the exhaustive instructions.

1. Add a Device:
   Tap '+' icon or "Add device" option within the MI Home app.
2. Select Device Type:
   Choose the device type accordingly, such as MI Home security camera, bedside lamp, Xiaomi smart speaker, and so on.
3. Connect to WiFi:

Table 1 IoT device specifications

Product model	Wireless connection	Work with MIJIA
Xiaomi motion sensor 2S	Bluetooth 5.0	Yes
Mi flood detector	Bluetooth 5.0	Yes
Xiaomi wireless switch	Bluetooth 5.0 BLE	Yes
Door and window sensor	Bluetooth 5.1 BLE	Yes
MI home security camera	Wi-fi IEEE 802.11 b/g/n 2.4GHz	Yes
Xiaomi smart speaker	Wi-fi IEEE 802.11 b/g/n 2.4GHz	Yes
Bedside lamp	Wi-fi IEEE 802.11 b/g/n 2.4GHz, Bluetooth 4.2 BLE	Yes

Follow the on-screen instruction to connect iPhone with the same wireless network to which smart devices that need to relate to.
4. Enter Pairing Mode:
   The pairing process varies with different equipment. Generally, it is prompted to press and hold a specific button or follow the manufacturer's instruction to put the device into pairing mode. And there will be a Light-Emitting Diode (LED) indicator on the device that blinks to present pairing is in progress.
5. Connect to Device:
   Follow the in-app prompts to connect the device with WiFi network credentials.
6. Profile Customize:
   After successfully pairing, we can rename the device per needs or just keep the default name. It could be also prompted to update the firmware if applicable.
7. Complete Setup:
   Once the setup is finished, all the IoT devices should be accessible with the MI Home app. For details, refer Fig. 3.

8.2 Intelligent Automation Scene

In this section, several smart home scenarios will be demonstrated using IoT devices. To begin, navigate to the automation feature in the MI Home app and select the plus button located at the top right corner to customize automation settings. Essentially, when creating an automation, three fields need to be configured. The first field pertains to "when the following situation occurs," which functions as the trigger condition. Subsequently, actions can be added under the "adjust the device to" attribute. Finally, the effective time frame will be defined according to the specific requirements of various scenarios. Furthermore, each of these three fields supports several specific actions. For example, when setting a trigger condition, options include device, timer, arrive/leave, and outdoor environment. For adjusting the device, choices encompass device, delay, notifications, and smart functionalities. As for the effective period, multiple time frames can be selected, such as 24 h cycles, daytime, nighttime, and so forth. The specific configuration is as follows:

1. Turn the lights on or off when a door or window is opened or closed. To accomplish this, select the door and window sensor and set its status to "on." Similarly, add the same condition and set its status to "off." Then, choose the bedside lamp to turn on or off accordingly. The effective time can be set to 24 h. In this scenario, when the door is detected as open or closed, the bedside lamp will execute the corresponding action.
2. Activate the camera upon detecting any movement. Utilize Xiaomi motion sensors to detect motion. Then, select the Mi Home security camera and enable the "turn on home surveillance" option, followed by setting it to 24 h operation for security purposes. In this case, if the homeowner is not at home, the motion

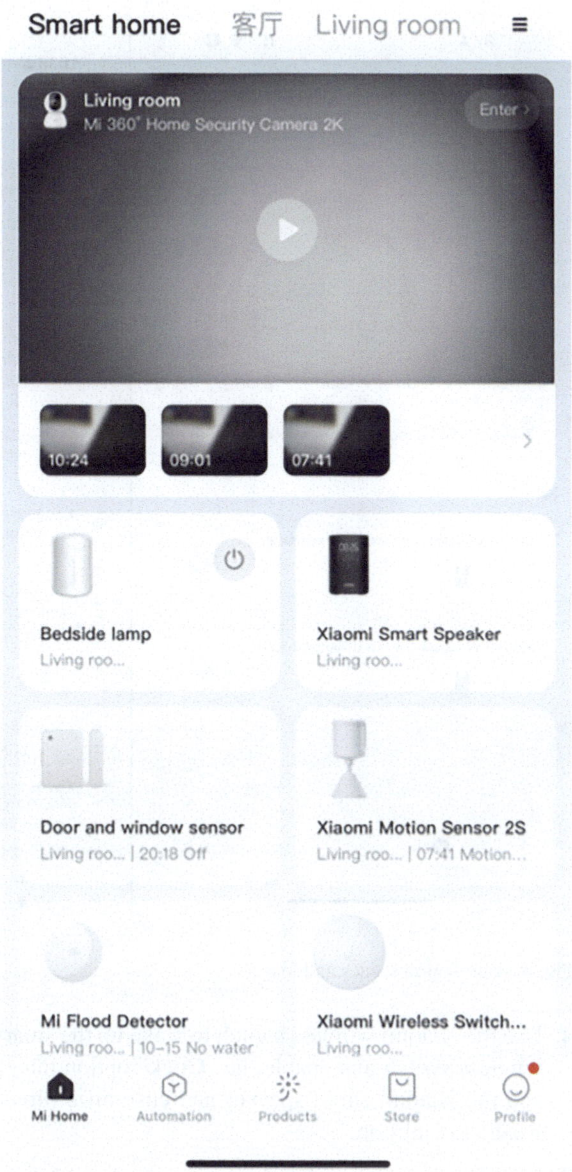

Fig. 3 MI Home applications

sensors can work in conjunction with the security camera to monitor any suspicious activity.

3. Address potential water leakage issues and send alarms to the Xiaomi smart speaker. Opt for the Mi flood detector and enable the "water leak notification" feature, which is provided by default. Then, configure the Xiaomi smart speaker to play music as an alarm. Set the duration to 24 h.

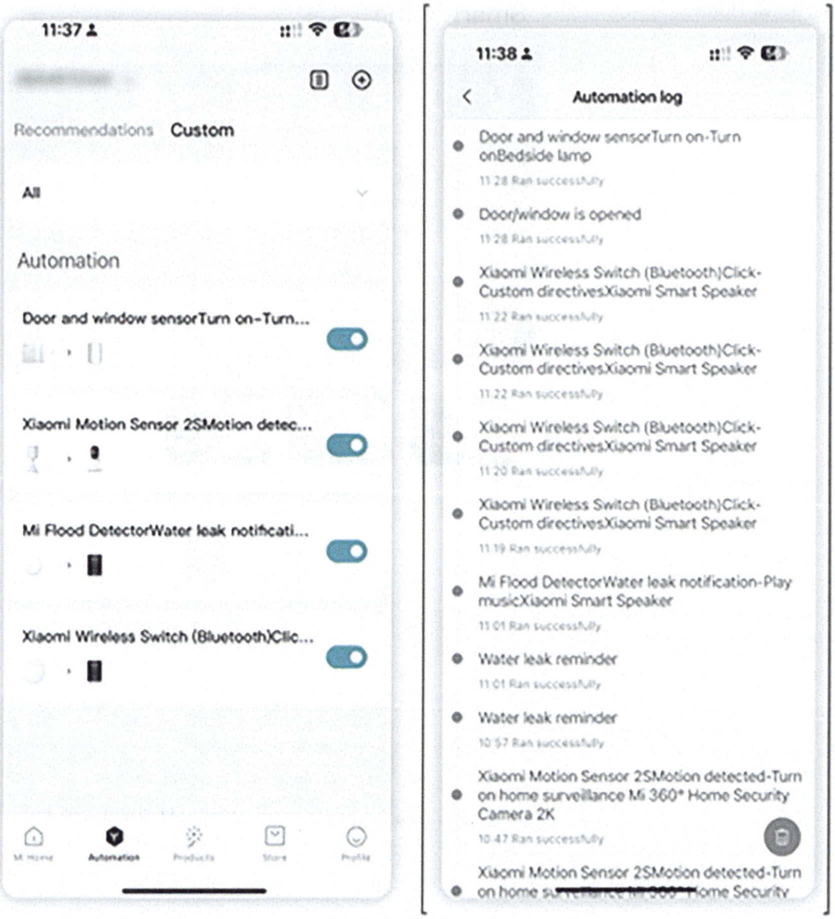

Fig. 4 Automation scenes and logs

4. Use the Xiaomi wireless switch to wake up the smart speaker. Select the Xiaomi wireless switch and enable the "Click" option integrated with the switch. Next, add the Xiaomi smart speaker and customize directives, such as "play today's weather condition."

Refer to Fig. 4 for customized scenes and related logs.

8.3 Analysis Based on Network Monitor Tool

As previously introduced, NMAP was employed to analyze network traffic and identify vulnerabilities. The study involved the installation of NMAP on a Mac OS platform and the subsequent execution of commands. Prior to commencing, the official website at Link was accessed to download the installer package for deploying the tool on a MacBook. Additionally, the NMAP Scripting Engine (NSP) was utilized to conduct vulnerability scans on devices connected within the intranet. As an integral component of NMAP, NSP stands out as one of the most versatile and robust features, enabling users to perform a wide array of network-related tasks (NMAP).

Furthermore, the script "nmap-vulners" was employed to demonstrate the detection of vulnerabilities from the Common Vulnerabilities and Exposures (CVE) database. CVE is utilized by security researchers to maintain a database cataloging and referencing individual vulnerabilities. The detailed procedure is as follows:

- Run the command nmap -PR < target IP address> to scan the IP address and open the port of IoT devices which connects to the intranet. After a while, the outcomes will return with specific information such as IP address, port status, and MAC address.
- Change to the directory of scripts by using the command cd /usr/share/nmap/scripts/.
- Then, clone the "nmap-vulners" GitHub repository by entering the following command in the terminal.
- git clone https://github.com/vulnersCom/nmap-vulners.git
- Finally, add the "—script" to the command of NMAP and execute the command to scan the vulnerabilities from the specific devices.
- nmap —script nmap-vulners -sV < target IP>.

Based on the results of the scans, no vulnerabilities were identified. This may be attributed to all the smart devices being up to date, implying that the manufacturer has potentially addressed security risks by upgrading their firmware and operating system versions. In this context, the attempt is to simulate a cyberattack on IoT devices to gain a better understanding of potential vulnerabilities.

8.4 DDoS Attack Simulation

Numerous varieties of cyberattacks have been documented in the past, and among these, Distributed Denial of Service (DDoS) attacks are widely recognized as one of the commonly employed methods by hackers due to their convenience [6]. These attacks are characterized by their relative ease of execution, often involving the acquisition of readily available toolkits and the launching of targeted assaults on digital systems. In this section, we will explore the utilization of a free DDoS tool known as Low Orbit Ion Cannon (LOIC) to conduct DDoS attacks on Internet-of-Things (IoT) equipment (How to DDoS). It is important to note that LOIC is incompatible with Apple chips like M1 and M2, leading us to opt for a Windows laptop with OS version 10 to create the necessary test environment for our experiments. In a study conducted by Ho et al. [11], which focused on various patterns associated with DDoS attacks, it was revealed that DDoS attacks is the preferred method that hackers often exploit when targeting IoT devices. As part of their research, they implemented a practical test environment involving various smart devices, such as Google Nest Hub Max, August Smart Lock Pro, Wyze Camera, and more. This experimental setup allowed them to assess the vulnerabilities of these devices to DDoS attacks. The research protocol commenced by employing Wireshark, a network protocol analyzer, to examine the regular traffic patterns within the communication channels connecting the smart devices. This initial phase of analysis provided valuable insights into the standard network traffic behavior in the absence of any external interference. Following this phase, the researchers introduced Slow Loris, a DDoS tool known for its effectiveness in overwhelming targeted systems. Slow Loris was used to initiate the first wave of attacks, which involved 60 terminals, simulating a real-world scenario of an impending DDoS attack. As the study progressed, the researchers escalated the intensity of their experiments by conducting a second wave of attacks. In this phase, they deployed 100 Slow Loris terminals to assess the IoT devices' resilience and response under a more substantial DDoS pressure. These experiments were designed to simulate the real-world challenges that IoT devices might face when subjected to a DDoS attack. The final phase of experimentation involved a combined assault using both the LOIC and Slow Loris tools. By combining these tools, the researchers aimed to understand the cumulative effects of two different DDoS attack vectors on IoT devices. This approach allowed them to examine the devices' responses and the intricacies of network traffic during a multifaceted DDoS attack scenario. To provide a comprehensive assessment of the experiments, the researchers utilized Wireshark to generate input/output graphics. These visual representations enabled them to elucidate the specific details of network traffic patterns during and after the DDoS attacks. The data gleaned from these graphics revealed that during the DDoS attacks, packet loss ranged from 8% to 10%. This demonstrated the considerable impact of the attacks on network stability and data transmission. Furthermore, the regular packet burst rate, which refers to the frequency of data packets being sent or received within a specific time frame, averaged 0.66 s per packet under normal network conditions. In stark contrast, the burst rate during the DDoS attacks averaged a significantly shorter duration of 0.28 s per packet. This finding highlighted a substantial increase in the rate at which data packets were transmitted during the DDoS attacks compared to the normal network traffic. The accelerated burst rate during the attacks indicated the severity of the disruption caused by the DDoS assaults on the network.

We intend to tailor our experimentation, taking inspiration from the test environment set up by the study mentioned above and optimize our procedures to address certain existing obstacles. One notable challenge is presented by Xiaomi smart devices, which can only connect via wireless means and do not support wired connections. This limitation complicates the exploration of normal network traffic between these devices. To overcome this, we plan to incorporate a Meraki Access Point (AP) into our experimental setup, which will serve as a bridge to connect both the WiFi network and the smart devices, facilitating the collection of relevant network data. The Meraki AP will play a pivotal role in the data collection process by receiving wireless data from IoT devices and transferring it into wired data. This transformation is possible because the AP itself is connected to the Internet via a cable, allowing it to bridge the gap between the wireless and wired network traffic. Additionally, Meraki AP comes equipped with a dedicated admin console that provides access to detailed logs and network activity information. This admin console will be instrumental in monitoring and recording the network traffic during our experiments. For a visual representation of our network setup, refer to Fig. 5, which illustrates the detailed network diagram, showcasing the role of Meraki Access Point in our experimental design.

With the preconfiguration in place, we proceeded to conduct a DDoS attack using LOIC. In this phase, we had three devices connected to Meraki Access Point (AP), each with its respective IP address. Our target devices were subject to the LOIC attack using their respective IP addresses. As we monitored the execution, it became evident that the bedside lamp fell victim to the attack. LOIC displayed 49,988 requests with the TCP method successfully executed, experiencing only 625 failed attempts. Interestingly, attempts to control the lamp via both the Mi Home application and Xiaomi speaker yielded no response. This indicated that the devices had indeed fallen prey to the DDoS attack, rendering them nonfunctional. After that, we extended the attacks to other devices, namely the security camera and Xiaomi speaker. However, these DDoS attacks were unsuccessful as LOIC failed to send the requests. It is plausible that these devices had protective mechanisms in place, thereby thwarting our DDoS attempts. Nonetheless, the situation provided us with the opportunity to capture network traffic data for analysis. Before initiating the

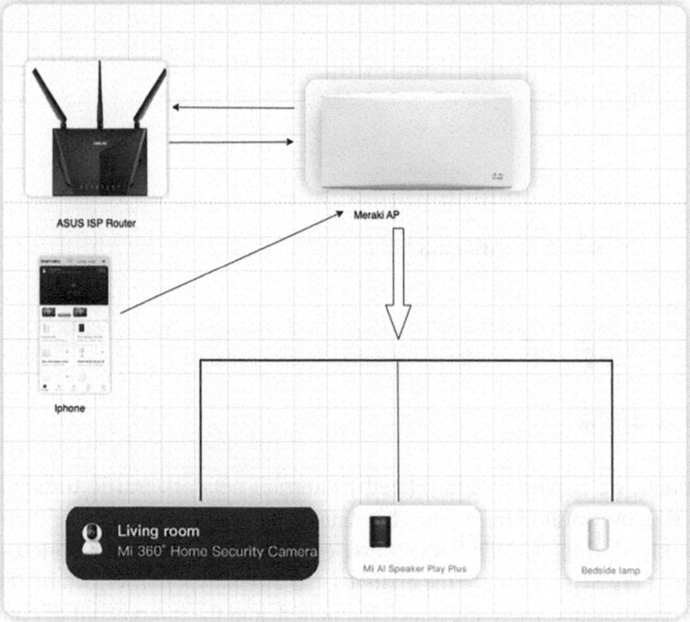

Fig. 5 Network diagram

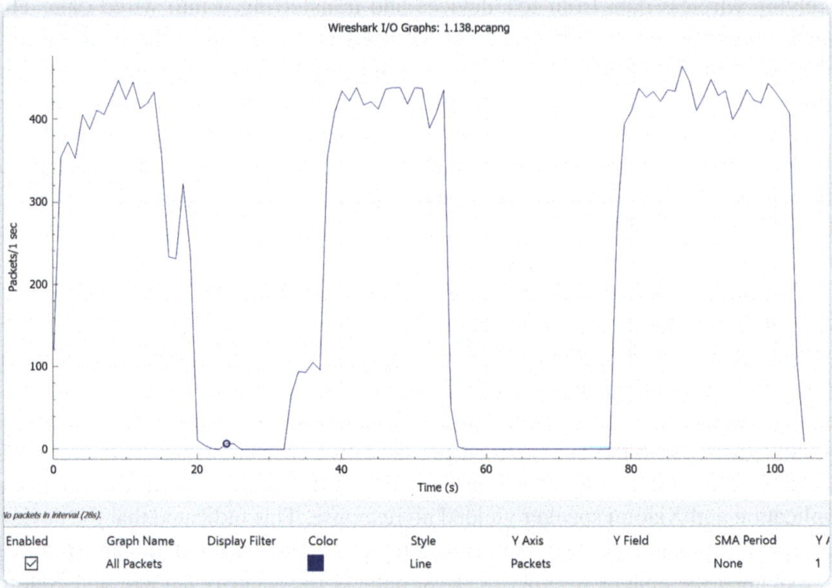

Fig. 6 Normal traffic

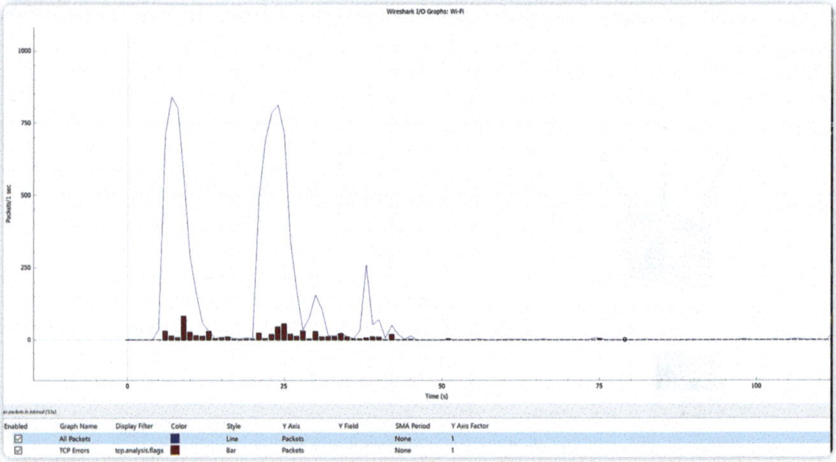

Fig. 7 Attacked traffic

DDoS attack, we captured data related to normal network traffic through the AP console, as demonstrated in Fig. 6. This data served as our baseline for comparison. Subsequently, during the DDoS attack, we continued to capture network traffic data (as shown in Fig. 7) to observe the changes and anomalies introduced by the attack. The TCP error means where the attack happened, and the lamp is out of control during these locations.

9 Survey on Social Engineering Attacks

In this chapter, a survey, deployed via Google Form, will be disseminated among individuals of varying occupations to shed light on the strategies for mitigating cyberattacks on IoT devices. The survey contains various sections, encapsulating aspects such as awareness of social engineering threats in IoT devices, personal experiences, security practices, and future apprehensions. With the data gathered from a multitude of questions, a comprehensive perspective regarding security apprehensions can be compiled. Concurrently, data from questions 4 and 5 in Fig. 8 indicate that the majority of participants have utilized IoT and smart devices, yet only half are aware of potential social engineering attacks. Pertaining to the instances of attacks targeting IoT devices, 66.7% of participants have not experienced an attack, 25% are unsure of previous attacks, and only 8.3% affirm encountering a

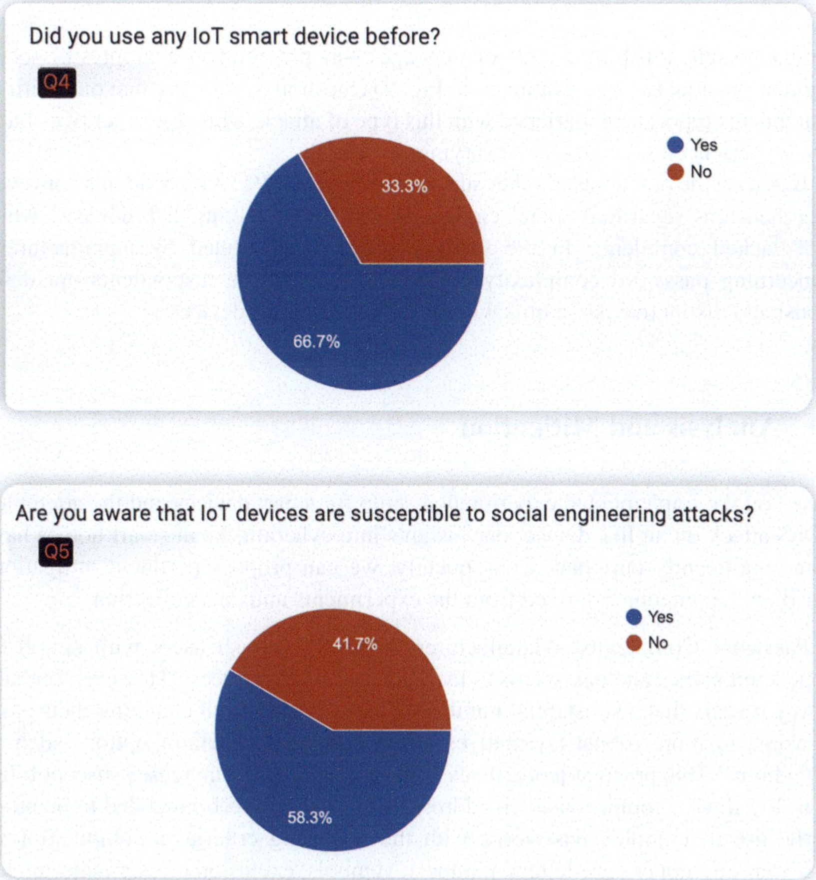

Fig. 8 Questions 4 and 5

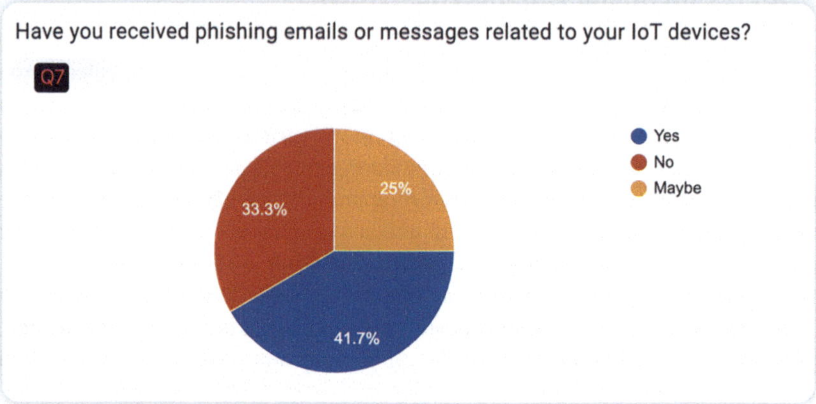

Fig. 9 Question 7

similar assault. Phishing emails or messages—as pre-eminent examples of social engineering attacks—are examined in Fig. 9 (Question 6) showing that one-third of respondents report no experience with this type of attack, while the other two-thirds have experienced such attacks to varying degrees.

Based on the outcomes of other survey inquiries, 41.7% of respondents conveyed apprehensions regarding social engineering attacks targeting IoT devices, while 25% lacked confidence in the security measures instituted by manufacturers. Concerning password complexity, fewer than half of the respondents opted for robust and distinctive passwords when configuring smart devices.

10 Analysis and Mitigation

Based on the implemented experimental setup for smart devices and the attempted DDoS attack on an IoT device, our insights into cyberattacks in smart homes have been significantly enriched. Consequently, we can propose pertinent mitigations based on the outcomes derived from the experiments and data collection.

- Password Complexity: Manufacturers typically furnish users with simplistic account names and passwords to facilitate easy device access. However, our survey reveals that a substantial number of users refrain from changing their passwords to more robust alternatives, often adhering to default options such as "admin." This practice jeopardizes their credentials and increases susceptibility to IoT device compromise. To address this issue, it is recommended to mandate the use of complex passwords with the following criteria: a combination of lower- and upper-case letters, numbers, symbols, exclusion of username-related elements, and a minimum length of eight characters. Implementing these password settings reduces the risk of brute force attacks.

- WiFi Security: Enhancing WiFi security poses a challenge due to the prevalence of smart devices relying on wireless connections. Additionally, the diversity of standards and protocols in wireless networks, compared to wired Internet, complicates the task [2]. To bolster wireless security, activating WiFi-Protected Access (WPA) or Wi-Fi-Protected Access 2 (WPA2) is recommended. These encryption methods safeguard data integrity and thwart denial-of-service attacks. Moreover, deactivating remote management ports from the router admin console prevents unauthorized remote-control attempts. Regular firmware updates, prompted by manufacturers, are crucial to addressing potential bugs and security vulnerabilities.
- Social Engineering Attacks: This distinctive form of attack targets the human element, focusing on smart device and smart homeowners rather than the physical devices themselves. Our survey underscored instances where participants fell victim to phishing emails or messages as part of social engineering attacks. To counter this threat, it is strongly advised not to click on any suspicious emails, particularly those containing links or requesting sensitive information and credentials. Exercising caution in this regard is pivotal for preventing social engineering breaches.
- Enhance Security Training: The survey results indicate that users still lack the understanding of security of IoT devices; this knowledge gap can be filled by providing additional training, for example, providing regular cybersecurity training, especially focusing on social engineering attacks and the security of IoT devices. This can cover aspects such as password management, the ability to identify spam emails, and recognizing false information and then conducting simulated social engineering attack scenarios, allowing users to experience potential risks firsthand. This helps deepen their understanding of threats and teaches practical response skills. Finally, regularly assess the effectiveness of training, collect user feedback, and adjust the training content based on the feedback received. This ensures that the training plan remains effective.

11 Conclusions

In concluding this chapter, we have delved deeply into the escalating security risks associated with the proliferation of IoT devices in smart home environments. This study has highlighted the increased vulnerability of these devices to a variety of cyberattacks, including denial of service, spoofing, man-in-the-middle, brute force, traffic analysis, and social engineering attacks. These threats pose significant risks to user privacy and data security, underlining the need for more robust protection measures. Our investigation has led to the identification and development of several key mitigation strategies. These include enhancing device security through firmware updates and stronger encryption, implementing safety assurance frameworks to standardize security practices across devices, and advocating for stronger authentication protocols. Particularly notable is the emphasis on user education and awareness, which is crucial in combating social engineering attacks and in fostering a more security-conscious user base.

The experimental component of this research has provided valuable insights into the practical application of these security measures. By simulating real-world scenarios, we have demonstrated the effectiveness of our proposed strategies in identifying and mitigating potential vulnerabilities. Looking ahead, the dynamic and ever-evolving nature of cybersecurity threats necessitates continuous research and adaptation of security measures. This study underscores the importance of ongoing research efforts in developing innovative, user-friendly security solutions tailored to the unique challenges posed by IoT devices in smart homes. By bridging the gap between theoretical research and practical applications, it contributes significantly to the field of cybersecurity, offering a roadmap for future developments in securing IoT ecosystems against emerging threats. Ultimately, our goal is to ensure that the advancements in smart home technology are matched by equally advanced security measures, allowing users to reap the benefits of IoT devices without compromising their safety and privacy.

12 Limitations

1. Due to the technology in smart devices' limitation, not all the equipment can be connected to wireless network which reduces the accuracy and makes them vulnerable to the DDoS attack.
2. Finding dedicated tools to perform brute force attack on all the devices which required user credential to login, it can help to conduct the solution or perform pentest in case that core IoT devices might be hacked and remotely controlled.

Acknowledgments This work is submitted to the University of Wollongong, Dubai as a requirement for a master thesis dissertation.

Appendix

Survey on Social Engineering Threats Targeting IoT Devices

References

1. Abdulbasit Darem, A. A. (March 2022). Cybersecurity threats and countermeasures of the smart home ecosystem. *IJCSNS International Journal of Computer Science and Network Security, 22*(3), 303–311.
2. Adam, M. M. E., & Abdallah, A. G. E. (2015). WIFI security. *International Journal of Advances in Engineering and Management (IJAEM), 2*(2), 143–149.

3. Akhilesh, R., Bills, O., Chilamkurti, N., & Chowdhury, M. J. M. (2022). Automated penetration testing framework for smart-home-based IoT devices. *Future Internet, 14*(10), 276. https://doi.org/10.3390/fi14100276
4. Al-Mutawa, R. F., & Eassa, F. (2020). A smart home system based on internet of things. *International Journal of Advanced Computer Science and Applications, 11*(2). https://doi.org/10.14569/ijacsa.2020.0110234
5. Batalla, J. M. (2019). Deployment of smart home management system at the edge: Mechanisms and protocols. *Neural Computing and Applications*, 1301–1315.
6. Cadzow, E. (2023). DDOS hackers using IoT devices to launch attacks. *Corero.* https://www.corero.com/ddos-hackers-using-iot-devices-to-launch-attacks/
7. Chatzisofroniou, G. (2016). *Getting the most out of evil twin.* https://www.census-labs.com/media/bsidesath2016-wifiphisher.pdf.
8. Chen, C., & Helal, S. (2011). *A device-centric approach to a safer internet of things*. https://doi.org/10.1145/2029932.2029934
9. Chifor, B. B. (2018). A security authorization scheme for smart home internet of things. *Future Generation Computer Systems*, 740–749.
10. Heartfield, R., Loukas, G., Budimir, S., Bezemskij, A., Fontaine, J., Filippoupolitis, A., & Roesch, E. B. (2018). A taxonomy of cyber-physical threats and impact in the smart home. *Computers & Security, 78*, 398–428. https://doi.org/10.1016/j.cose.2018.07.011
11. Ho, S., Greeson, H., & Karabiyik, U. (2022). *Smart home forensics: Identifying DDoS attack patterns on IoT devices.*
12. Jain, G. C., & Anubha. (2021). Application of SNORT and wireshark in network traffic analysis. *IOP Conference Series, 1119*(1), 012007. https://doi.org/10.1088/1757-899x/1119/1/012007
13. Kabir, S., Gope, P., & Mohanty, S. P. (2022). A security-enabled safety assurance framework for IoT-based smart homes. *IEEE Transactions on Industry Applications, 59*(1), 6–14. https://doi.org/10.1109/tia.2022.3176257
14. Khoei, T. T., Slimane, H. O., & Kaabouch, N. (2022). *A comprehensive survey on the cybersecurity of smart grids: Cyber-attacks, detection, countermeasure techniques, and future directions*. arXiv preprint arXiv:2207.07738.
15. Liao, S., Zhou, C., Zhao, Y., Zhang, Z., Zhang, C., Gao, Y., & Zhong, G. (2020). *A comprehensive detection approach of Nmap: Principles, rules and experiments*. https://doi.org/10.1109/cyberc49757.2020.00020
16. Lin, H., & Bergmann, N. W. (2016). Iot privacy and security challenges for smart home environments. *Information, 7*(3), 44.
17. Liu, Y. Z. (2016). *An efficient privacy protection solution for smart home application platform.* In 2016 2nd IEEE International conference on computer and communications (ICCC).
18. Machin, M., Guiochet, J., Waeselynck, H., Blanquart, J., Roy, M., & Masson, L. (2018). SMOF: A safety monitoring framework for autonomous systems. *IEEE Transactions on Systems, Man, and Cybernetics, 48*(5), 702–715. https://doi.org/10.1109/tsmc.2016.2633291
19. Mayes, K. E., Markantonakis, K., Francis, L., & Hancke, G. P. (2010). Nfc security threats. *Smart Card Technology International Magazine*, 42–47.
20. Pardeep Kumar, A. B. (2017, April). Anonymous secure framework in connected smart home environments. *IEEE Transactions on Information Forensics and Security.*
21. Rahmati, A., Fernandes, E., Eykholt, K., & Prakash, A. (2018). *Tyche: Risk-based permissions for smart home platforms*. arXiv preprint arXiv:1801.04609, pp. 1–15. https://www.arxiv.org/pdf/1801.04609.pdf
22. Satapathy, U. A. (2018). *An ECC based lightweight authentication protocol for mobile phone in smart home*. In IEEE 13th international conference on industrial and information systems (ICIIS).
23. Stamm, S., Ramzan, Z., & Jakobsson, M. (2006). Drive-by pharming. In *Proceedings of international conference on information and communications security* (pp. 495–506). Springer.

24. Tirumala, S. S., Nepal, N., & Ray, S. K. (2022). Raspberry Pi-based intelligent cyber defense systems for SMEs and smart-homes: An exploratory study. *EAI Endorsed Transactions on Smart Cities, 6*(18), e4. https://doi.org/10.4108/eetsc.v6i18.2345
25. Tu, N. T., Chengyu, S., Zhiyun, Q., Srikanth, V. K., Edward, J. M. C., & Patrick, M. (2018). IotSan fortifying the safety of IoT systems. *ACM Proceedings*, 191–203. https://jglobal.jst.go.jp/en/detail?JGLOBAL_ID=202002257278991866
26. Usman, M., Muthukkumarasamy, V., & Wu, X. (2015). Mobile agent-based cross-layer anomaly detection in smart home sensor networks using fuzzy logic. *IEEE Transactions on Consumer Electronics, 61*(2), 197–205. https://doi.org/10.1109/tce.2015.7150594
27. Visoottiviseth, V., Akarasiriwong, P., Chaiyasart, S., & Chotivatunyu, S. (2017). *PENTOS: Penetration testing tool for Internet of Thing devices*. https://doi.org/10.1109/tencon.2017.8228241.

Security Challenges and Solutions in IoT: Analyzing Threats, Architectures, and Policies

Yashika Jamwal, Priti Kumari, Ashutosh Mishra, Nilesh Goel, and Mrinal Goswami

1 Introduction

Since the early 90 s, technology has advanced rapidly, making modern computing systems like laptops and phones more efficient, powerful, and common in households. By 2025, it is estimated that around 18.22 billion [1] devices will be in use worldwide. However, this proliferation has also led to an increase in cyberattacks, with about 2200 attacks occurring daily, affecting over 800,000 people each day [2]. Cybersecurity is essential to protect these systems from adversaries and prevent harm to corporations, entities, and users.

While traditional systems have robust security mechanisms, the Internet of Things (IoT) poses unique challenges. The IoT is a rapidly growing industry where low computational power limits the use of traditional security measures [3]. A survey by a leading antivirus company reported a 30% increase in IoT malware attacks in 2020, affecting 32.4 million devices globally [2]. A significant example is the Mirai Botnet attack in October 2016, which caused Internet outages and impacted major social media networks and OTT platforms. The potential risks extend beyond corporations, affecting sectors like health care and agriculture, where compromised IoT devices could endanger lives. This underscores the critical need for robust security in IoT devices [4].

The dynamic and low-power nature of IoT devices prevents a standardized security architecture, making them vulnerable to various attacks [3]. To address these challenges, understanding the current research on IoT device architecture, attacks, and security solutions is crucial.

Y. Jamwal · P. Kumari · A. Mishra · N. Goel (✉)
BITS Pilani Dubai Campus, Dubai, UAE
e-mail: nileshgoel@dubai.bits-pilani.ac.in

M. Goswami
Assam Down Town University, Guwahati, Assam, India

© The Author(s), under exclusive license to Springer Nature Switzerland AG 2025

A. Mishra et al. (eds.), *Hardware Security: Challenges and Solutions*,
https://doi.org/10.1007/978-3-031-81213-2_10

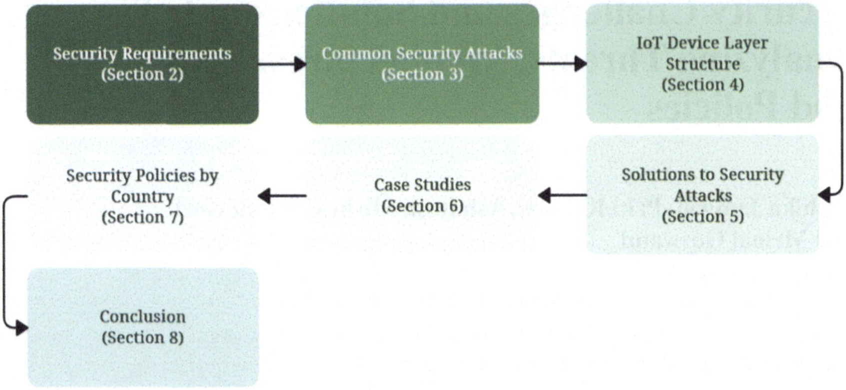

Fig. 1 Chapter organization

This chapter addresses these issues in several ways:

- It organizes information from state-of-the-art research on security requirements, attacks, IoT layers, and security solutions, making it easier for readers to understand how these aspects of security interrelate, with a focus on hardware.
- Case studies highlight the importance of securing IoT devices across various domains and identify which layers and requirements are most affected by attacks, as well as the appropriate security solutions.
- The chapter also presents information on security policies from leading countries, helping readers understand the global importance of protecting IoT devices.

The rest of the chapter is structured as follows: Sect. 2 delves into security requirements, Sect. 3 lists common security attacks targeting IoT hardware, Sect. 4 discusses attempts to structure IoT device layers, Sect. 5 covers solutions to the security attacks detailed in Sects. 6 and 7 presents case studies linking all security aspects, Sect. 7 reviews security policies from different countries, and Sect. 8 concludes the chapter (Fig. 1).

2 Security Requirements

Manufacturers should not expect consumers to grasp the complexities of IoT security [5–7]. Consumers often prioritize technology use without understanding its security. This is particularly concerning for IoT systems, where security is currently poor, making security requirements crucial [8–17]. These requirements offer insights into necessary security services and help tailor security measures for specific devices. They can be classified into six broad categories: Trust, Privacy, Identity Management, Integrity, Availability, and Confidentiality, further detailed as follows:

1. **Trust**: Trust in IoT involves protecting user data, verifying identity, and building confidence in third parties. Given the nonstandardized nature of IoT devices, establishing trust is challenging but essential for user confidence [5–7, 16, 18]. Trust includes security, which assures users that the system is safe, and privacy, which governs the disclosure of data, thereby fostering trust between users and third parties [18]. According to [18], trust properties are classified into five categories:
   (a) Objective properties of trustees (e.g., security, dependability).
   (b) Subjective properties of trustees (e.g., honesty).
   (c) Subjective properties of trustors (e.g., trust willingness).
   (d) Objective properties of trustors (e.g., trust criteria).
   (e) Context or environment of operation.

   In accordance with the suggestions given by [18], to have a trustworthy IoT system, the following goals must be achieved:

   (a) **Layerwise Goals**: These goals are pertinent to the specific layers present in the IoT architecture, which generally include the perception, network, and application layer (Refer to **Sect. 4** for further information on different layers present in IoT devices).
   (b) **Cross-Layer Goals**: These goals can be applied to all of the layers present in the IoT architecture. Advantages of trust mechanisms include certainty in collaboration, flexibility to adapt to security changes, and potential uniformity across IoT devices using trust chains. For a trustworthy IoT system, goals should include reliable data sensing, secure data transmission, and user-friendly services, with **trust management** and **establishment** being key components [5, 6].

2. **Privacy**: In IoT, privacy is critical due to the vast amount of data generated. It ensures user data is protected from unauthorized use, identity is verified, and users are not traceable or linkable [5, 6, 16, 19–22]. Privacy includes.

   (a) **Anonymity**: Keeping user information hidden from unauthorized sources [22].
   (b) **Resilience**: Ensuring security even during attacks or failures to prevent data leakage [22].

3. **Identity Management**: Due to the sheer number of IoT devices and the relationship between them, identity management plays an important role in ensuring that unauthorized users or untrusted devices have the appropriate action taken against them [6]. Identity management can be further classified into three categories:

   (a) **Identification**: Due to the presence of many devices in an IoT network, it is important for the smart device to know when to and when not to reveal its identity. The devices must be capable of identifying other qualified devices and users to distinguish them from threats as well as provide strong security with system restrictions [6, 12, 21].

(b) **Authentication/Authenticity**: Authentication involves recognizing the identity of a user or device with specific credentials. It ensures that the device has rights to access the network, and the network has rights to connect to the device [22]. Authenticity gives the assurance of information and transactions being genuine by validating the parties participating in the process. It ensures that only the authorized parties can be involved in the communication [7, 11, 20, 21, 23]. **Authorization** involves giving consent to entities for performing actions such as giving permissions to users, devices, or applications [17, 21, 23]. **Access control** ensures which entities can access which aspect of the network [21].
(c) **Accountability**: It helps trace the actions and events of a particular user or system to establish responsibility for their respective actions [13, 16]. **Auditability** is the ability of a system to conduct continuous monitoring of actions performed by the system and the users [16].

4. **Integrity**: Integrity ensures data is not altered, fabricated, or misused by unauthorized parties, using methods like checksums and cyclic redundancy checks [6, 7, 12, 16, 18–20, 22–24]. It includes.

   (a) **Nonrepudiation**: It is the ability of a system to confirm the occurrence/nonoccurrence of actions performed by users or the system [16, 23].
   (b) **Authenticity**: Ensuring only authorized users can access resources, preventing data tampering [11].

5. **Availability**: Ensures that IoT services are accessible to authorized users or devices, even during catastrophes, ensuring proper operation of computing systems [6, 7, 11, 16, 18–20, 23, 24]. It involves.

   (a) **Scalability**: The system's ability to adjust to changes in processing demands [25].
   (b) **Interoperability**: Seamless communication and coordination between various IoT systems and devices [25].
   (c) **Self-organization**: It determines how capable an IoT system is in adjusting itself to remain operational even if there is failure in the system due to malfunctioning or attack [22].

   For a device to always be available, **resilience** must be incorporated. The device must be capable of ensuring security and availability even in the case of attacks and failures (prevention of data leakage) [22].

6. **Confidentiality**: Protects sensitive data from unauthorized access during exchanges between users and devices. Confidentiality is maintained through access control, authentication, and encryption, with user anonymity playing a key role [7, 11, 13, 16–24]. It includes.

   (a) **Information Security Management**: It is a way for an organization to protect sensitive data from various threats and vulnerabilities [26].
   (b) **Key Agreement**: A protocol used in cryptography to ensure both parties agree on encryption keys [26].

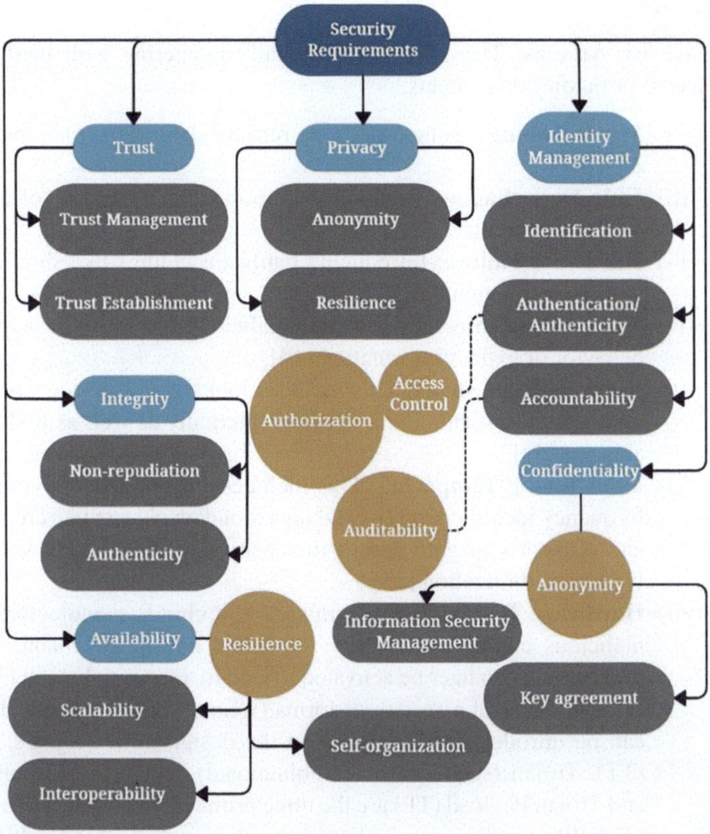

Fig. 2 Classification of security requirements

Confidentiality requires ensuring information is not shared with unauthorized parties and maintaining user anonymity [22] (Fig. 2).

3 Security Attacks

Security requirements help us understand what services need to secure sensitive information. However, understanding how these requirements can be affected by attacks is important to secure infrastructure and prevent loss of sensitive information. This section provides insights into common hardware [5, 27] and encryption [9, 12, 13, 27–30] attacks.

1. **Hardware**.
   (a) **Invasive Attacks**: These involve physically tampering with hardware to access or modify its contents.
      (i) **Microprobing**: Using a probe to retrieve data or instructions from a chip [31].
      (ii) **Chip Depackaging**: Removing the protective packaging of an IC to access its circuitry.
      (iii) **Hardware Failure**: Introducing hardware failures like short circuits to disrupt functionality [32].
      (iv) **Malign Data Injection**: Injecting malicious data into a device to alter behavior or extract information [14].
      (v) **Reverse Engineering**: This includes looking at a device's design in order to understand how it operates internally as well as find vulnerabilities [33].
      (vi) **Tag Cloning/Tampering**: This includes changing or copying radio-frequency identification (RFID) tags found on objects like credit cards and passports to gain access to sensitive information or safe areas without authorization [34].
      (vii) **Hardware Trojan (HT)**: In this case, a chip is manufactured with malicious coding or circuitry implanted during the manufacturing process that can later be activated to lead to a security breach [34–36]. HT can occur in a variety of forms (weaponized to passive HT) and can be introduced at any stage in the design flow. Trojan's Trigger (TT), Trojan (sequence or combinational) Triggering Circuit (TC), and Trojan Payload (TP) are the three primary components of a Trojan [37]. When a Trojan is activated, it releases its payload, which may cause the circuit to become completely or partly dysfunctional, cause a denial of service, change the circuit's characteristics (e.g., by aging factors), or leak confidential data.
      (viii) **Physical Damage**: Physically damaging the device to disrupt its operation or gain information [34].
      (ix) **Outage Attack**: Disrupting power or communication channels to cause an outage [34].
      (x) **Sleep Deprivation**: Disrupting sleep mode to extract data [7, 36].
      (xi) **Battery Drainage Attack**: Draining the battery to force shutdown or make the device unavailable [14].
      (xii) **Malicious Code Injection**: Injecting harmful code to access sensitive information and enable other attacks [6, 14, 34, 38, 39].
      (xiii) **Node Cloning**: Replicating or forging devices in a network to use for unauthorized purposes [14, 32].
      (xiv) **Social Engineering**: Attackers can physically interact with the users involved with an IoT system [34].

(b) **Semi-invasive attacks**: These involve tampering with the device without direct electrical contact with the chip. These attacks do not require complete removal of the hardware components of the device [32, 40, 41].

  (i) **Fault Injection**: Inducing abnormal behavior using X-rays, electromagnetic fields, etc. [32, 40].

(c) **Noninvasive Attacks**: These types of attacks do not need physical access to the target device. Instead, they exploit weaknesses in hardware implementation such as signal leakage or electromagnetic radiation to collect data from the device or alter its behavior.

  (i) **Jamming Attacks**: Transmitting signals to disrupt IoT communication, causing denial-of-service or unauthorized access [10, 36].
  (ii) **Object Jamming**: Targeting a specific device by overwhelming its communication [34].
  (iii) **Node Jamming**: Disrupting communication between nodes in an IoT network [7].
  (iv) **RFID Jamming**: In this type of attack, the attacker deliberately sends a powerful radio signal to interfere with the transmission between an RFID tag and a reader, causing the tag to become useless. This may cause difficulties with security, inventory management, and other RFID-based systems.
  (v) **Eavesdropping**: Extracting information from a communication channel without detection, enabling further attacks. Usually, an existing IoT device is used to collect and misuse confidential information [4–6, 12, 14, 16, 34, 42].
  (vi) **Side Channel Attack**: The state of the CPUs memory cache can be leaked using side channel attacks. This can reveal memory access patterns and hence can be used for cryptanalysis. Ref. [44] provides insights into the cached attacks and their countermeasures with respect to AES [35].
  (vii) **Unauthorized Access**: In this kind of attack, the attacker gains access to the device or network without proper authorization and can then carry out harmful activities such as stealing data, manipulating data, or even device takeover [34].
  (viii) **Node Capture**: In this attack, a network node is physically taken over and gains access to its data [42].
  (ix) **Object Replication**: This type of attack involves attacker creating a fake device to gain unauthorized control [34].
  (x) **Camouflage**: The technique of hiding malicious nodes or devices in the network by passing them off as genuine ones. The purpose of this attack is to achieve unapproved access to a target device or system and avoid detection by security controls [34].
  (xi) **Sinkhole**: This type of attack involves adversaries capturing control of one or more network nodes and reroutes traffic to them instead of

the target node This attack may be used to capture or modify data, launch other crimes, or both [6, 7, 20, 34, 36, 45].

2. **Encryption.**

Cryptography uses mathematical techniques to protect data from unauthorized access [46]. Lightweight cryptography addresses constraints in resource-limited devices [22], balancing security with complexity. Encryption attacks target software, hardware, data, and protocols.

(a) **Cryptanalysis**: It is the process in which an adversary tries to find a weakness in the cryptographic algorithms to decipher the ciphertext, usually without knowing the cipher key. It is not limited to the algorithm and can be in how the algorithm has been applied [20, 47–49].

(i) **Side-channel**: This attack involves gaining encryption keys by applying the following attacks [16, 27, 34]:

1. **Timing Analysis**: Attacker can gain sensitive information by observing how long the system takes to respond to cryptography algorithms [42].
2. **Power Analysis**: Power consumption of a cryptographic device is observed to find patterns related to the algorithm [10]. Ref. [50] provides a comprehensive survey on the various kinds of power analysis attack techniques.
3. **Fault Analysis**: These attacks force devices to work in nonideal conditions such as high temperatures, glitches in power supply, etc., especially when cryptographic operations are being performed in the system [51].
4. **Electromagnetic Analysis**: An attacker can use unintentional electromagnetic emissions for eavesdropping on operations/data handling of the device [52].
5. **Caching Attack**: The state of the CPU's memory cache can be leaked using side-channel attacks. This can reveal memory access patterns and hence can be used for cryptanalysis. Ref. [44] provides insights into the cached attacks and their countermeasures with respect to AES.

(b) **Key Management Attacks**: Key management involves generating, distributing, and supporting the code keys. This process is susceptible to a number of passive and active attacks [8, 35, 53]. This process is also explained in depth in Ref. [23]. The article provides a comprehensive view on dynamic and static key management for wireless sensors (Fig. 3).

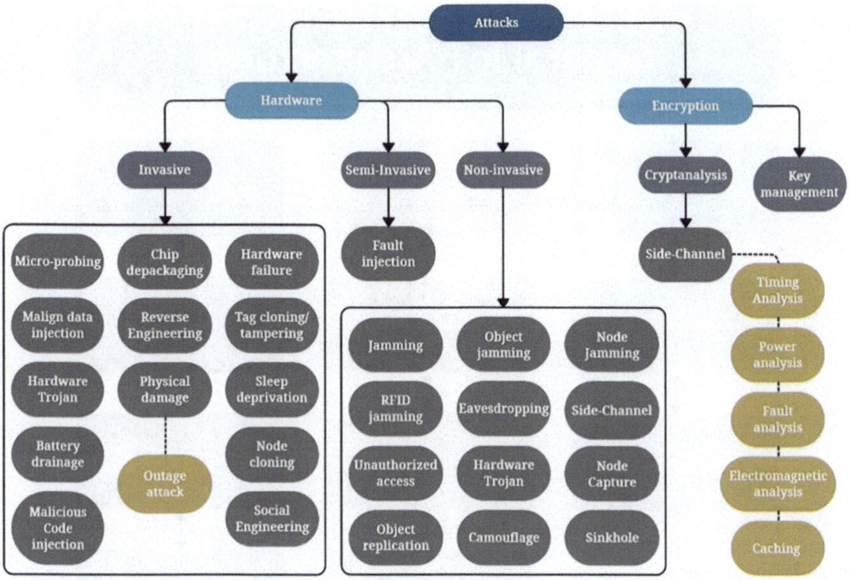

Fig. 3 Classification of security attacks related to hardware and hardware-based encryption

4 Common IoT Layers

- **IoT Architecture and Its Protocol/Technology**
- IoT architecture refers to the complex arrangement of elements that make up IoT networking systems, including sensors, actuators, cloud services, protocols, and layers. In general, it is separated into layers that enable administrators to assess, keep an eye on, and uphold the integrity of the system. There is not a single, universal IoT design that researchers throughout the world agree upon. Some researchers claim that the IoT architecture has three layers, whereas others prefer it to have four. Because of advancements in the Internet of Things, they contend that the three-layer architecture cannot satisfy application requirements. Due to the problem with security and privacy in IoT, five- and six-layer designs have also been proposed. The Internet of Things' security and privacy needs may be met by a recently proposed design, according to experts [34, 54]. This section provides an insight into the different architectures proposed, with an emphasis on the perception layer and common technologies associated with it [34, 42, 54, 55] (Fig. 4).
- **Three-Layer Architecture.**

  (a) **Perception/Sensing Layer/Physical Layer**: This layer is the core of standard IoT architecture, primarily responsible for facilitating and securing data transmission between the application and perception layers [56]. It gathers data from the physical environment (e.g., temperature and light) using vari-

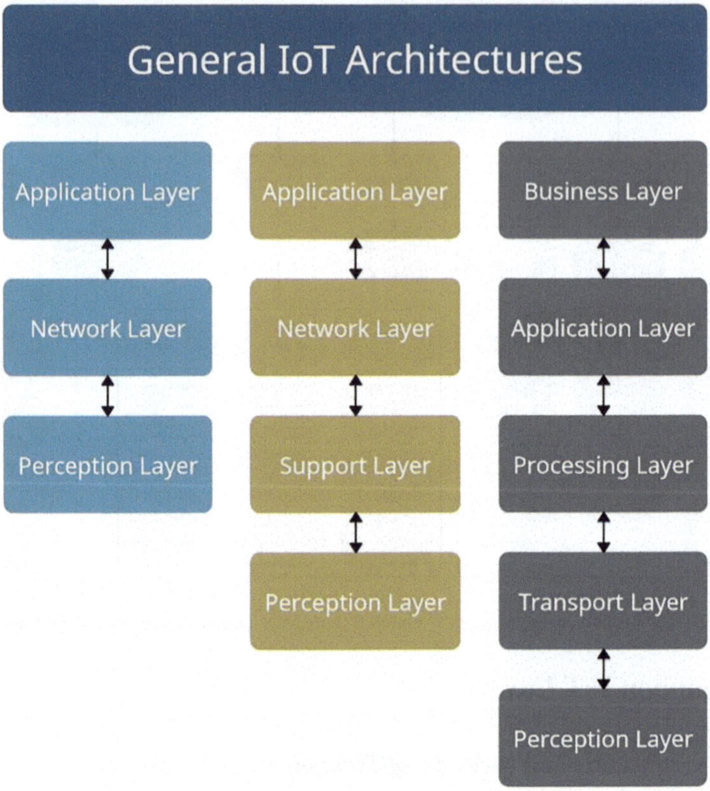

Fig. 4 General architectures proposed (three, four, and five layers)

ous sensing technologies like sensors, cameras, and GPS terminals, typically through short-range communication [57]. The layer is divided into two parts [58]: the **perception node**, which includes sensors and controllers, and the **perception network**, which links the network layer via short- and long-range communication technologies. Technologies in this layer include RFID, ZigBee, sensor nodes, sensor gateways, and electronic data interface (EDI).

(i) **Perception Node**:

1. **RFID**: Enables contactless communication and automatic identification of items using radio waves. RFID tags consist of a transceiver/antenna and a microchip, activated by an RFID reader. RFID technologies are categorized by [34, 42, 59]:

   (a) **Operating Frequency**: Low (LF), high (HF), ultra-high (UHF), and microwave.
   (b) **Power Supply**: Active, passive, and battery-assisted passive tags.
   (c) **Memory Type**: Read-only, write-once, and read-many.

2. **WSN**: An IoT Wireless Sensor Network (WSN) is a collection of spatially scattered sensors that gather and wirelessly transmit data on environmental factors like temperature, sound, and humidity [42].
3. **RS-232**: A common protocol for serial communication that connects a computer to its peripherals.

(ii) **Perception Network.**

1. **Short-Range Wireless Technology**.
   (a) **Bluetooth**: Used to exchange data between nearby devices, originally designed to replace wired connections [34, 42].
   (b) **NFC**: Enables wireless information exchange by touching or close proximity, useful for transactions, data sharing, and access control [34].
   (c) **Zigbee**: A wireless communication technology based on IEEE 802.15.4, enabling low-power devices to communicate over radio and create a Wireless Personal Area Network (WPAN) [22, 24, 34, 42].
   (d) **Wi-Fi**: Based on IEEE 802.11, Wi-Fi enables the creation of WLANs, with current networks offering up to 6.75 Gbps over distances up to 382 km [34, 42, 59].
   (e) **BLE**: A low-power variant of Bluetooth focused on IoT, with two sub-specifications: BLE single mode and BLE dual mode, incompatible with classic Bluetooth devices [22, 24, 34, 59].
   (f) **Z-WAVE**: A short-distance communication standard that creates mesh networks for device communication, with a range of about 30 m.
   (g) **Light Fidelity**: A wireless protocol that transmits data using Light-Emitting Diodes (LEDs) in the visible range, offering data rates 100 times faster than Wi-Fi.

2. **Long-Range Communication Technology**.
   (a) **Narrow Band (NB) IoT**: Long-Term Evolution (LTE)-based Low-Power Wide-Area Network (LPWAN) technology for wide-area coverage, offering low-power, low-cost communication with up to 10 years of device lifespan [60]. This technology eliminates the need for gateways and is particularly versatile with the 2G, 3G, and 4G technologies [60].
   (b) **SigFox**: LPWAN communication technology using Ultra-Narrow Band (UNB) modulation to reduce energy usage and enhance receiver sensitivity [60].
   (c) **LoRaWAN**: LPWAN technology using LoRa spread spectrum modulation to wirelessly connect battery-operated devices [22, 60].

(d) **5G (Fifth Generation) and Beyond**: Fifth-generation cellular network offering high data rates (up to 20 Gbps), low latency (1 ms), and seamless connectivity across various environments using technologies like mmWave and Massive MIMO [42, 60].

(b) **Network Layer**: The typical IoT architecture‘s network layer is its most advanced layer. The core layer of IoT manages data forwarding and network integration. It ensures device connectivity and provides addressing and routing capabilities. Key protocols include CORPL, RPL, CARP, 6LowPAN, and 6TiSCH [7, 14, 22, 34].

(c) **Transport Layer**: This layer focuses on end-to-end communication, offering reliability and sequencing of data packets. It includes Transmission Control Protocol (TCP) (for reliable connections), User Datagram Protocol (UDP) (for less reliable applications), and Stream Control Transmission Protocol (SCTP) (for multistreaming and multihoming). Security protocols like Data Transport Layer Security (DTLS) and Transport Layer Security (TLS) are commonly used for UDP and TCP, respectively [10, 57, 62, 63]. Time-Synchronized Mesh Protocol (TSMP) aids in synchronizing devices in wireless networks.

(d) **Application Layer**: As the top layer, it bridges users and apps, offering customized services for various applications like disaster monitoring and health tracking. Protocols include CoAP, XMPP, MQTT, DDS, AMQP, and REST [7, 10, 22, 25, 36, 59, 61].

- **Four-Layer Architecture.**
- To address the limitations of the three-layer architecture, a four-layer model adds a support layer [42, 55]. This layer handles security by verifying that data is sent by authorized users and ensuring its safety before it reaches the network layer.
- **Five-Layer Architecture.**

The five-layer architecture builds on the four-layer model to address additional security and storage issues. It includes the perception, transport, and application layers, plus two new layers: the processing layer (which processes and filters data) and the business layer (which manages IoT applications, business models, and user privacy) [32] (Table 1).

5 Security Solutions

The security attacks discussed earlier pose serious threats to IoT devices, disrupting businesses and compromising user privacy. Understanding how to mitigate these attacks is crucial. This section provides a comprehensive list of solutions relevant to the attacks mentioned. For more details on each attack and solutions, refer to the following works:

Table 1 Common architecture layers with respective technologies/protocols

Layer	Technology/Protocol
Application layer	CoAP, MQTT, DDS, HTTP, AMQP, XMPP, DNP,PTP, STOMP, REST [62]
Transport layer	TCP/UDP, TLS/SSL, UDP/IP, DTLS, TSMP, SCTP
Network layer	IPV4/IPV6, RPL, 6LoWPAN, CoRPL, CARP
Data link layer	Short Range: NFC, ZigBee, Bluetooth, Weightless, BLE, Wireless HART, ENOcean, Z-Wave, Digimesh, THREAD Long Range: WIFI, 5G, LoRaWAN, 4G/LTE, Symphony link, Sigfox, e-MTC, BB-IOT, EGGSM, GSM,IOT, Ethernet, Satellite, PLC, NB-IoT, Wi-Fi direct.
Sensing layer	RFID, Sensor Nodes, Sensor Gateways, WSN, Modbus, RS-232, USB, QR Code

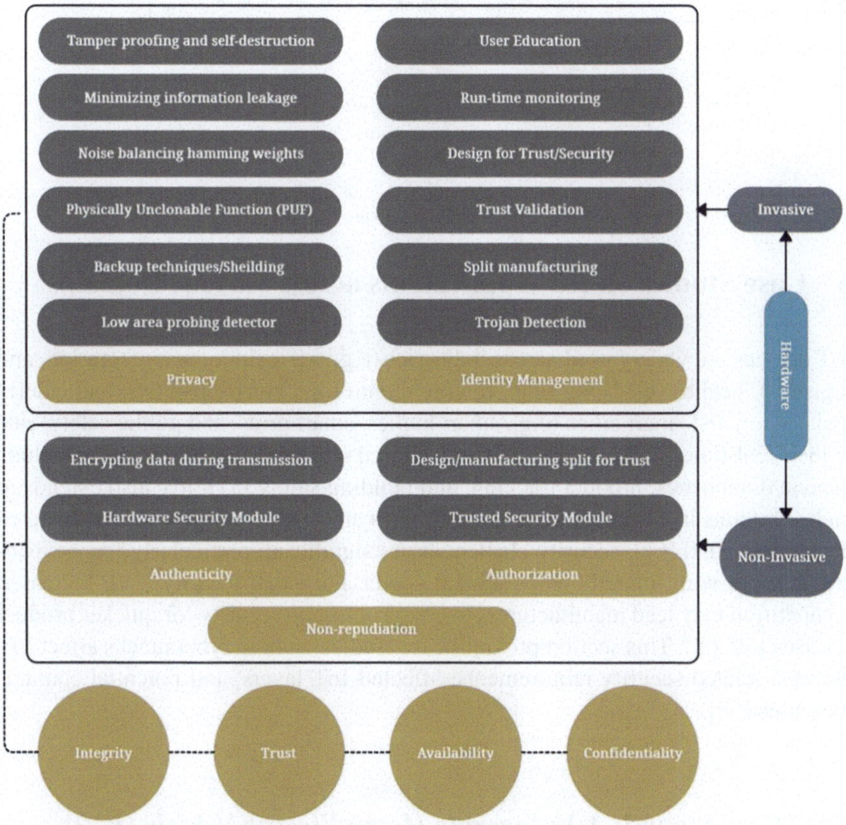

Fig. 5 Common solutions and requirements for hardware attacks

- Noninvasive: [6, 7, 10, 20, 34–36, 42, 43, 45, 64, 65].
- Invasive: [14, 31, 32, 33–35, 37, 64, 65].
- Cryptanalytic [20, 27, 52, 66].
- Key management: [53] (Figs. 5 and 6).

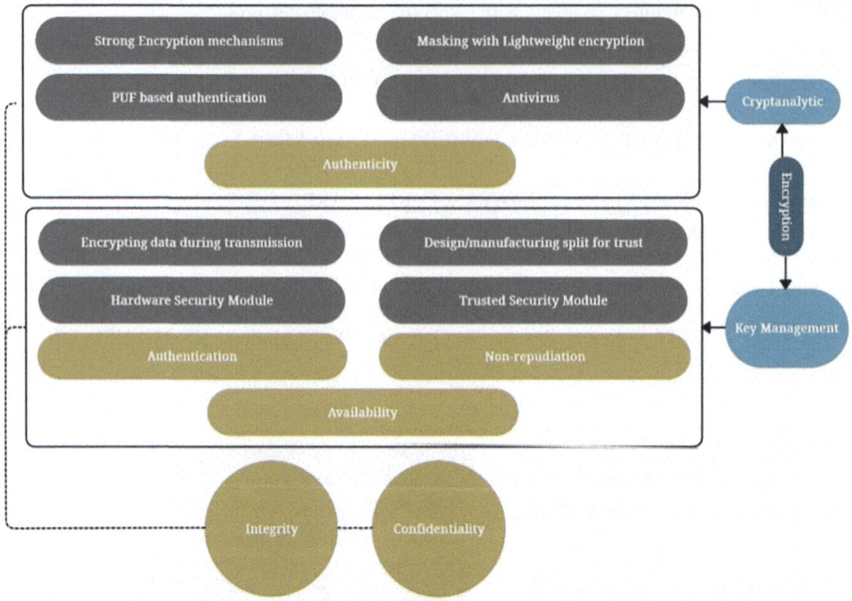

Fig. 6 Common solutions and requirements for hardware-based encryption attacks

6 Case Studies on IoT Applications across Various Domains

IoT devices are widely used across industries (e.g., agriculture, manufacturing, and logistics), health well-being (e.g., remote monitoring, smart hospitals, and elderly assistance), and smart cities (e.g., smart homes, smart grids, and public safety) due to their real-time situational sensing and tailored services [57]. Applications include vehicle diagnostics, product tracking, and building safety [57]. IoT also extends to niche domains like military settings [39]. With an estimated 24.1 billion connected devices by 2030 [67], security challenges are significant, particularly for inexpensive devices with limited computational power and weak security [39]. Increased competition may lead manufacturers to compromise on security for quicker product releases [39, 68]. This section provides case studies on how cyberattacks affect IoT devices, related security requirements, affected IoT layers, and potential countermeasures [57].

6.1 Case Study I: A Vulnerable Home Electric Vehicle (EV) Charging Station

Smart grids, in the Smart City domain, are intelligent systems that deliver and monitor energy, providing efficient charging for electric vehicles. These stations, usually connected to the grid, allow for a fast, safe, and easy charging process [57]. An extension of these charging stations are home electric vehicle charging stations which are designed for use in areas such as homes or private parking spaces. Public charging stations have a robust power grid management system, which home electric vehicle charging stations usually do not require since they do not deal with any of the advanced features such as a payment machine. Hence, when manufacturers add additional features to these stations, there is always a possibility of attackers exploiting vulnerabilities, primarily due to lack of security testing [69].

Security researchers [69] discovered a vulnerability in the firmware of ChargePoint, Inc's EV charger. This physical layer attack enabled further exploits at the software and data link layers. Through code injection during the password authentication phase, attackers could bypass login credentials, alter system code, cause buffer overflows leading to DoS attacks, and access the system kernel for file manipulation, potentially disrupting the device or electricity supply [39]. These attacks compromise key security requirements like authorization, authentication, availability, and confidentiality. Recommended countermeasures include static validation (e.g., using strncpy() instead of strcpy()), parameter validation (e.g., privilege level checks), and dynamic verification techniques (e.g., memory corruption detection).

6.2 Case Study II: Thermostat with the Capability to Spy

In the Smart City domain, smart homes/buildings use devices and sensors for security (e.g., video surveillance), service automation (e.g., irrigation), and entertainment [57]. These systems are popular for their convenience in remote control and scheduling, saving time, energy, and money [70]. An example is Google's Nest Thermostat, which optimizes power usage by tracking user habits and storing data in the Nest Cloud, including system and software logs [71]. Researchers in [71] found a vulnerability in the Nest Thermostat's physical layer, where pressing the hard-reset button puts it into Device Firmware Update (DFU) mode, allowing attackers to insert malicious code via USB. This could turn the thermostat into a botnet client, compromising the network and other connected devices, leading to privacy breaches, unauthorized surveillance, and data theft. Key security requirements like privacy, authentication, availability, confidentiality, integrity, and trust are compromised. Mitigation strategies include implementing secure boot features, integrity checks on software images, and securing API endpoints [39, 71].

6.3 Case Study III: Vulnerabilities in a Pacemaker

The healthcare sector, part of the health well-being domain, uses sensing devices like Implantable Medical Devices (IMDs) to monitor patient vitals (e.g., heart rate, temperature, cholesterol, and sugar levels). Data from these devices, shared via communication technologies like BLE, ZigBee, and Wireless HART, helps caretakers diagnose and monitor health, while patients manage their conditions through web/mobile apps [57]. The rise of healthcare IoT has introduced “human device networks,” increasing the risk of attacks [72].

One such device is a “pacemaker,” which sends electrical impulses to the heart whenever there is an anomaly in the regular heartbeat (correcting slow heartbeat or no heartbeat) [73, 74] and is now equipped with wireless technology for monitoring. However, these advancements also raise security concerns due to their proximity to vital organs. In 2017, MedSec Holdings Ltd. identified vulnerabilities in Abbott Laboratories’ pacemakers, affecting 465,000 devices [60, 72, 74]. These weaknesses, mainly in the communication layer [75], allowed unauthorized access and malicious RF commands, leading to battery drainage and potential life-threatening malfunctions [72–74]. The security issues impacted trust, authorization, integrity, and availability. A firmware update was released to limit “RF wake-up” commands and secure data transmission [73, 74].

6.4 Case Study IV: Stuxnet—The World’s First Military Cyberweapon

Military Internet of Things (MIoT) integrates IoT technologies into Defense Technology and Military operations, with applications in communications, logistics, military medicine, and weapon systems [76–78]. Security attacks, particularly from nation-state actors [79] or advanced persistent threats [80], are major concerns due to potential catastrophic impact on both military personnel and civilians [81]. The 2010 Stuxnet worm attack, known as “the world’s first digital weapon”, targeted Iran’s nuclear facility, specifically the Industrial Control Systems (ICS) used to operate the uranium enrichment centrifuges [82–84]. The worm exploited zero-day vulnerabilities in Windows, spread via USB sticks, and evaded detection by using a digital certificate. Once the worm gained control over the systems, the centrifuges were spined till failure [83]. This attack highlighted the importance of introducing stronger security measures in ICS such as mandatory encryption and better protection of critical infrastructure [4]. Confidentiality, integrity, availability, authentication, authorization, nonrepudiation, and accountability were some of the security requirements affected by this attack (Table 2).

7 Country-Specific Policies for Securing IoT Devices

Referring to the previous sections in the chapter, it is imperative that having a legal framework to safeguard IoT data and its users from security breaches is of utmost importance. It is the responsibility of authorities to enact laws that deal directly with IoT devices and prevent any unwanted repercussions. In this section, some of the laws and standards enacted by different nations on IoT security are discussed [4].

Table 2 Summary of the presented case studies

Aspect	Case study I: home EV charging station	Case study II: thermostat	Case study III: pacemaker vulnerabilities	Case study IV: stuxnet
Domain	Smart grids/ smart city	Smart home/ building	Health/Well-being	Military IoT
Device	Home EV charger	Thermostat	Pacemaker	Industrial control systems
Affected layers	Physical, software, and data-link layers	Physical layer	Communication layer (can lead to effects on physical layer)	Physical, communication, and application layers
Attack	Code injection, buffer overflow	Malicious code injection, botnet creation	Improper authentication scheme, battery drainage attack	Zero-day vulnerabilities, digital certificate evasion
Effect of attack	Unauthorized access, DoS, file modification	Network vulnerability	Disrupted functions, unauthorized access, loss of life	Centrifuge failure, ICS disruption
Security requirements	Authorization, authentication, availability, confidentiality	Privacy, authentication, availability, confidentiality, integrity, trust	Trust, authorization, integrity, availability	Confidentiality, accountability, integrity, availability, authorization, non-repudiation
Mitigation	Static and parameter validation, dynamic verification	Secure boot, integrity checks, API security	Firmware update, RF command limitation, encryption	Mandatory encryption, improved physical and system security

1. **European Union (EU)**: EU's General Data Protection Regulation (GDPR) governs the use, processing, and storage of personal data, including data from IoT devices. It enforces rights such as the right to be forgotten, data transparency, and user authorization [85]. However, implementing GDPR principles in IoT devices is challenging due to the lack of standardized architecture and protocols [4, 85, 86].
2. **Japan:** Japan's data protection policy, the "Act on the Protection of Personal Information" (APPI), mandates consent before collecting or using personal information, restricts cross-border data transfers, and allows individuals to view and update their information. The APPI, revised in 2017 and further amended in 2018, introduced stricter rules and fines for data breaches, particularly ahead of the 2020 Tokyo Olympics [87, 88].
3. **United Arab Emirates (UAE):** On March 22, 2018, the UAE introduced its IoT Regulatory Policy, with data privacy laws enforced through the Abu Dhabi Global Market (ADGM) and the Dubai International Financial Centre (DIFC), based on global norms. The policy includes rules on data reduction, storage lim-

its, and high encryption standards. The Personal Data Protection Law (PDPL) [89], enforced on January 2, 2022, is part of UAE's efforts to lead in IoT regulation in the Middle East [90, 91].

4. **Australia:** Australia's "Code of Practice," released on September 3, 2020, outlines 13 rules for manufacturers on smart product security. Developed after national consultations, it aims to enhance IoT security. The Australian Signals Directorate also provides guidelines on safe IoT usage. Australia's main data protection law, The Privacy Act 1988, has undergone numerous amendments [89, 91–94].
5. **The United States of America (US):** [95, 96] The US addresses IoT through various laws, such as the "Federal Privacy Act" of 1974, HIPAA (1996), COPPA (1998), and ECPA (1986), though these were not specifically designed for IoT. California's SB 327, effective January 1, 2020, was the first IoT-specific law, followed by Oregon's House Bill 2395. NIST issued guidelines in June 2019 to minimize IoT cybersecurity and privacy risks. The IoT Cybersecurity Improvement Act of 2019 and UL's IoT security rating system also aim to enhance IoT security, though their effectiveness is debated [4] (Fig. 7).

8 Conclusions

The rapid advancement of technology has led to a surge in connected devices, increasing their vulnerability to cyberattacks. IoT devices are particularly at risk due to their low computational power, lack of standardized security mechanisms, and the broad threat landscape they face. Addressing these vulnerabilities requires focused research on IoT architecture, potential attacks, and effective security solutions. Security in IoT is complex, and manufacturers should not assume that consumers fully understand these complexities. A multifaceted approach is needed to tackle key security requirements such as trust, privacy, identity management, integrity, availability, and confidentiality. These elements are crucial for protecting sensitive information and ensuring reliable IoT operations. Understanding these security requirements helps manufacturers focus on critical areas during device production and in developing mitigation strategies. Identifying and analyzing specific attacks that IoT devices are vulnerable to is equally important. This knowledge allows for the creation of more effective security strategies. A thorough understanding of hardware-specific attacks, including invasive, semi-invasive, and noninvasive types, can enhance device protection and prevent data loss. Staying updated on current and emerging attack techniques helps security professionals address new vulnerabilities effectively. The complexity of IoT devices is further increased by the need to understand various IoT architectures and their specific layers. Each layer introduces potential vulnerabilities during data transmission or processing. Studying these layers aids in recognizing evolving security needs and their role in preventing attacks. It is also crucial to understand how different security solutions are implemented across IoT architecture layers. This knowledge supports the development of comprehensive security strategies, optimizes resource use, and ensures best practices.

Fig. 7 Summary of prominent security policies in different countries

Case studies illustrate the diverse applications of IoT devices and highlight unique vulnerabilities in each. These examples emphasize the importance of comprehensive security strategies that address varied attack types across different IoT applications. Additionally, understanding security policies provides insights into data regulation, protection, innovation, and consumer trust. It ensures IoT devices meet security standards set by different countries. However, many countries, especially in Asia, the Middle East, Latin America, and Africa, still lack robust IoT security laws. As the demand for IoT devices grows, so will the number of adversaries. Implementing proper security measures before product release is crucial to prevent exploitation. This chapter underscores the importance of understanding these critical topics to enhance IoT security.

References

1. https://www.statista.com/statistics/245501/multiple-mobile-device-ownership-worldwide/
2. https://us.norton.com/internetsecurity-emerging-threats-cyberthreat-trends-cybersecurity-threat-review.html
3. Vallabhaneni, R. (2024). *Effects of data breaches on Internet of Things (IoT) devices within the proliferation of daily-life integrated devices.*
4. Kerr, P. K., Rollins, J., & Theohary, C. A. (2010). *The Stuxnet computer Worm: Harbinger of an emerging warfare capability.* https://digital.library.unt.edu/ark:/67531/metadc31393/m1/1/high_res_d/R41524_2010Dec09.pdf
5. Alaba, F. A., Othman, M., Hashem, I. A. T., & Alotaibi, F. (2017). Internet of things security: A survey. *Journal of Network and Computer Applications, 88*, 10–28. https://doi.org/10.1016/j.jnca.2017.04.002
6. Ogonji, M. M., Okeyo, G., & Wafula, J. M. (2020). A survey on privacy and security of internet of things. *Computer Science Review, 38*, 100312. https://doi.org/10.1016/j.cosrev.2020.100312
7. Yugha, R., & Chithra, S. (2020). A survey on technologies and security protocols: Reference for future generation IoT. *Journal of Network and Computer Applications, 169*, 102763. https://doi.org/10.1016/j.jnca.2020.102763
8. Yousefpoor, M. S., & Barati, H. (2018). Dynamic key management algorithms in wireless sensor networks: A survey. *Computer Communications*. https://doi.org/10.1016/j.comcom.2018.11.005
9. Abiodun, O. I., Abiodun, E. O., Alawida, M., Alkhawaldeh, R. S., & Arshad, H. (2021). A review on the security of the internet of things: Challenges and solutions. *Wireless Personal Communications, 119*(3), 2603–2637. https://doi.org/10.1007/s11277-021-08348-9
10. Mohanta, B. K., Jena, D., Satapathy, U., & Patnaik, S. (2020). Survey on IoT security:Challenges and solution using machine learning, artificial intelligence and blockchain technology. *Internet of Things*, 100227. https://doi.org/10.1016/j.iot.2020.100227
11. Chettri, L., & Bera, R. (2019). A comprehensive survey on Internet of Things (IoT) towards 5G wireless systems. *IEEE Internet of Things Journal*, 1–1. https://doi.org/10.1109/jiot.2019.2948888.
12. Tahsien, S. M., Karimipour, H., & Spachos, P. (2020). Machine learning based solutions for security of internet of things (IoT): A survey. *Journal of Network and Computer Applications*, 102630. https://doi.org/10.1016/j.jnca.2020.102630
13. Neshenko, N., Bou-Harb, E., Crichigno, J., Kaddoum, G., & Ghani, N. (2019). Demystifying IoT security: An exhaustive survey on IoT vulnerabilities and a first empirical look on internet-scale IoT exploitations. *IEEE Communications Surveys & Tutorials, 21*(3), 2702–2733. https://doi.org/10.1109/comst.2019.2910750
14. Iqbal, W., Abbas, H., Daneshmand, M., Rauf, B., & Abbas, Y. (2020). An in-depth analysis of IoT security requirements, challenges and their countermeasures via software defined security. *IEEE Internet of Things Journal*, 1–1. https://doi.org/10.1109/jiot.2020.2997651.
15. Di Martino, B., Rak, M., Ficco, M., Esposito, A., Maisto, S. A., & Nacchia, S. (2018). Internet of things reference architectures, security and interoperability: A survey. *Internet of Things, 1–2*, 99–112. https://doi.org/10.1016/j.iot.2018.08.008
16. Mosenia, A., & Jha, N. K. (2017). A comprehensive study of security of internet-of-things. *IEEE Transactions on Emerging Topics in Computing, 5*(4), 586–602. https://doi.org/10.1109/tetc.2016.2606384
17. Khatua, P. K., Ramachandaramurthy, V. K., Kasinathan, P., Yong, J. Y., Pasupuleti, J., & Rajagopalan, A. (2019). Application and assessment of internet of things toward the sustainability of energy systems: Challenges and issues. *Sustainable Cities and Society*, 101957. https://doi.org/10.1016/j.scs.2019.101957
18. Frustaci, M., Pace, P., Aloi, G., & Fortino, G. (2018). Evaluating critical security issues of the IoT world: Present and future challenges. *IEEE Internet of Things Journal, 5*(4), 2483–2495. https://doi.org/10.1109/jiot.2017.2767291

19. Kouicem, D. E., Bouabdallah, A., & Lakhlef, H. (2018). Internet of things security: A top-down survey. *Computer Networks, 141*, 199–221. https://doi.org/10.1016/j.comnet.2018.03.012
20. Radoglou Grammatikis, P. I., Sarigiannidis, P. G., & Moscholios, I. D. (2018). Securing the Internet of Things: Challenges, threats and solutions. *Internet of Things*. https://doi.org/10.1016/j.iot.2018.11.003.
21. Omolara, A. E., Alabdulatif, A., Abiodun, O. I., Alawida, M., Alabdulatif, A., Alshoura, W. H., & Arshad, H. (2022). The internet of things security: A survey encompassing unexplored areas and new insights. *Computers & Security, 112*, 102494. https://doi.org/10.1016/j.cose.2021.102494
22. Meneghello, F., Calore, M., Zucchetto, D., Polese, M., & Zanella, A. (2019). IoT: Internet of threats? A survey of practical security vulnerabilities in real IoT devices. *IEEE Internet of Things Journal, 6*(5), 8182–8201. https://doi.org/10.1109/jiot.2019.2935189
23. Raskar, C., & Nema, S. (2019). *A review on Internet of things*. In 2019 International conference on intelligent sustainable systems (ICISS). https://doi.org/10.1109/iss1.2019.8908105.
24. Lounis, K., & Zulkernine, M. (2020). Attacks and defenses in short-range wireless technologies for IoT. *IEEE Access*, 1–1. https://doi.org/10.1109/access.2020.2993553.
25. Hassan, R., Qamar, F., Hasan, M. K., Aman, A. H. M., & Ahmed, A. S. (2020). Internet of things and its applications: A comprehensive survey. *Symmetry, 12*(10), 1674. https://doi.org/10.3390/sym12101674
26. Yang, W., Wang, S., Sahri, N. M., Karie, N. M., Ahmed, M., & Valli, C. (2021). Biometrics for internet-of-things security: A review. *Sensors, 21*(18), 6163. https://doi.org/10.3390/s21186163
27. Sengupta, J., Ruj, S., & Bit, S. D. (2019). A comprehensive survey on attacks, security issues and blockchain solutions for IoT and IIoT. *Journal of Network and Computer Applications*, 102481. https://doi.org/10.1016/j.jnca.2019.102481
28. Ahmad, R., & Alsmadi, I. (2021). Machine learning approaches to IoT security: A systematic literature review. *Internet of Things, 14*, 100365. https://doi.org/10.1016/j.iot.2021.100365
29. Stoyanova, M., Nikoloudakis, Y., Panagiotakis, S., Pallis, E., & Markakis, E. K. (2020). A survey on the Internet of Things (IoT) forensics: Challenges, approaches and open issues. *IEEE Communications Surveys & Tutorials*, 1–1. https://doi.org/10.1109/comst.2019.2962586.
30. Rachit, B. S., & Ragiri, P. R. (2021). Security trends in internet of things: A survey. *SN Applied Sciences, 3*(1). https://doi.org/10.1007/s42452-021-04156-9
31. Skorobogatov, S. (2017). *How microprobing can attack encrypted memory*. In 2017 Euromicro conference on digital system design (DSD). https://doi.org/10.1109/dsd.2017.69.
32. Makhdoom, I., Abolhasan, M., Lipman, J., Liu, R. P., & Ni, W. (2018). Anatomy of threats to The Internet of Things. *IEEE Communications Surveys & Tutorials*, 1–1. https://doi.org/10.1109/comst.2018.2874978.
33. Shwartz, O., Mathov, Y., Bohadana, M., Oren, Y., & Elovici, Y. (2018). Reverse engineering IoT devices: Effective techniques and methods. *IEEE Internet of Things Journal*, 1–1. https://doi.org/10.1109/jiot.2018.2875240.
34. Akram, H., Konstantas, D., & Mahyoub, M. (2018). A comprehensive IoT attacks survey based on a building-blocked reference model. *International Journal of Advanced Computer Science and Applications, 9*(3). https://doi.org/10.14569/ijacsa.2018.090349
35. Sidhu, S., Mohd, B. J., & Hayajneh, T. (2019). Hardware security in IoT devices with emphasis on hardware trojans. *Journal of Sensor and Actuator Networks, 8*(3), 42. https://doi.org/10.3390/jsan8030042
36. Defense Department. (2005, January 1). *Defense science board task force on high performance microchip supply*. Government. Defense Department. https://www.govinfo.gov/app/details/GOVPUB-D-PURL-LPS66631
37. Vakil, A., Mirzaeian, A., Homayoun, H., Karimi, N., & Sasan, A. (2021). AVATAR: NN-assisted variation aware timing analysis and reporting for hardware trojan detection. *IEEE Access, 9*, 92881–92900. https://doi.org/10.1109/access.2021.3093160

38. Khan, M. A., & Salah, K. (2018). IoT security: Review, blockchain solutions, and open challenges. *Future Generation Computer Systems, 82*, 395–411. https://doi.org/10.1016/j.future.2017.11.022
39. Alladi, T., Chamola, V., Sikdar, B., & Choo, K.-K. R. (2020). Consumer IoT: Security vulnerability case studies and solutions. *IEEE Consumer Electronics Magazine, 9*(2), 17–25. https://doi.org/10.1109/mce.2019.2953740
40. Tudosa, I., Picariello, F., Balestrieri, E., De Vito, L., & Lamonaca, F. (2019). *Hardware security in IoT era: The role of measurements and instrumentation*. In 2019 II workshop on metrology for industry 4.0 and IoT (MetroInd4.0&IoT). https://doi.org/10.1109/metroi4.2019.8792895.
41. Bhunia, S., & Tehranipoor, M. (2019). Physical attacks and countermeasures. In *Elsevier eBooks* (pp. 245–290). https://doi.org/10.1016/b978-0-12-812477-2.00015-0.
42. Burhan, M., Rehman, R., Khan, B., & Kim, B.-S. (2018). IoT elements, layered architectures and security issues: A comprehensive survey. *Sensors, 18*(9), 2796. https://doi.org/10.3390/s18092796
43. Panchal, A. C., Khadse, V. M., & Mahalle, P. N. (2018). *Security issues in IIoT: A comprehensive survey of attacks on IIoT and its countermeasures*. In 2018 IEEE global conference on wireless computing and networking (GCWCN). https://doi.org/10.1109/gcwcn.2018.8668630.
44. Osvik, D. A., Shamir, A., & Tromer, E. (2006). Cache attacks and countermeasures: The case of AES. In *Lecture notes in computer science* (pp. 1–20). https://doi.org/10.1007/11605805_1.
45. Noor, M. b. M., & Hassan, W. H. (2018). Current research on internet of things (IoT) security: A survey. *Computer Networks*. https://doi.org/10.1016/j.comnet.2018.11.025
46. https://www.nist.gov/cryptography
47. Ray, S., Mishra, K. N., & Dutta, S. (2021). Big data security issues from the perspective of IoT and cloud computing: A review. *Recent Advances in Computer Science and Communications, 14*(7), 2057–2078. https://doi.org/10.2174/2666255813666200224092717
48. https://owasp.org/www-community/attacks/Cryptanalysis
49. Shibuya, K., & Issiki, S. (2014). Evaluation of metallic mold surfaces polished by an industrial robot with stick whetstones. *International Journal of Automation Technology, 8*(2), 253–263. https://doi.org/10.20965/ijat.2014.p0253
50. Randolph, M., & Diehl, W. (2020). Power side-channel attack analysis: A review of 20 years of study for the layman. *Cryptography, 4*(2), 15. https://doi.org/10.3390/cryptography4020015
51. Baksi, A., Bhasin, S., Breier, J., Jap, D., & Saha, D. (2022). A survey on fault attacks on symmetric key cryptosystems. *ACM Computing Surveys, 55*(4), 1–34. https://doi.org/10.1145/3530054
52. Sayakkara, A., Le-Khac, N., & Scanlon, M. (2019). A survey of electromagnetic side-channel attacks and discussion on their case-progressing potential for digital forensics. *Digital Investigation, 29*, 43–54. https://doi.org/10.1016/j.diin.2019.03.002
53. Cabrera-Gutierrez, A. J., Castillo, E., Escobar-Molero, A., Alvarez-Bermejo, J. A., Morales, D. P., & Parrilla, L. (2022). Integration of hardware security modules and permissioned blockchain in industrial IoT networks. *IEEE Access, 10*, 114331–114345. https://doi.org/10.1109/access.2022.3217815
54. Altulaihan, E., Almaiah, M. A., & Aljughaiman, A. (2022). Cybersecurity threats, countermeasures and mitigation techniques on the IoT: Future research directions. *Electronics, 11*(20), 3330. https://doi.org/10.3390/electronics11203330
55. Da Xu, L., He, W., & Li, S. (2014). Internet of things in industries: A survey. *IEEE Transactions on Industrial Informatics, 10*(4), 2233–2243. https://doi.org/10.1109/tii.2014.2300753
56. Kolandaisamy, R., Subaramaniam, K., & Jalil, A. B. (2021). *A study on comprehensive risk level analysis of IoT attacks*. In 2021 International conference on artificial intelligence and smart systems (ICAIS). https://doi.org/10.1109/icais50930.2021.9395858.
57. Borgia, E. (2014). The internet of things vision: Key features, applications and open issues. *Computer Communications, 54*, 1–31. https://doi.org/10.1016/j.comcom.2014.09.008

58. Gupta, B., & Quamara, M. (2018). An overview of internet of things (IoT): Architectural aspects, challenges, and protocols. *Concurrency and Computation Practice and Experience, 32*(21). https://doi.org/10.1002/cpe.4946
59. Mansour, M., Gamal, A., Ahmed, A. I., Said, L. A., Elbaz, A., Herencsar, N., & Soltan, A. (2023). Internet of things: A comprehensive overview on protocols, architectures, technologies, simulation tools, and future directions. *Energies, 16*(8), 3465. https://doi.org/10.3390/en16083465
60. https://www.medsec.com/industry-involvement
61. Yudidharma, A., Nathaniel, N., Gimli, T. N., Achmad, S., & Kurniawan, A. (2023). A systematic literature review: Messaging protocols and electronic platforms used in the internet of things for the purpose of building smart homes. *Procedia Computer Science, 216*, 194–203. https://doi.org/10.1016/j.procs.2022.12.127
62. Tiburski, R. T., Amaral, L. A., de Matos, E., de Azevedo, D. F. G., & Hessel, F. (2017). *Evaluating the use of TLS and DTLS protocols in IoT middleware systems applied to E-health.* In 2017 14th IEEE annual consumer communications & networking conference (CCNC). https://doi.org/10.1109/ccnc.2017.7983155.
63. Čolaković, A., & Hadžialić, M. (2018). Internet of things (IoT): A review of enabling technologies, challenges, and open research issues. *Computer Networks, 144*, 17–39. https://doi.org/10.1016/j.comnet.2018.07.017
64. Rooney, C., Seeam, A., & Bellekens, X. (2018). Creation and detection of hardware trojans using non-invasive off-the-shelf technologies. *Electronics, 7*(7), 124. https://doi.org/10.3390/electronics7070124
65. Balaji, S., Nathani, K., & Santhakumar, R. (2019). IoT technology, applications and challenges: A contemporary survey. *Wireless Personal Communications, 108*(1), 363–388. https://doi.org/10.1007/s11277-019-06407-w
66. Sustek, L. (2011). Hardware security module. In *Springer eBooks* (pp. 535–538). https://doi.org/10.1007/978-1-4419-5906-5_509.
67. Hou, K. M., Diao, X., Shi, H., Ding, H., Zhou, H., & De Vaulx, C. (2023b). Trends and challenges in AIoT/IIOT/IoT implementation. *Sensors, 23*(11), 5074. https://doi.org/10.3390/s23115074
68. Sharma, R., & Arya, R. (2023, November). Security threats and measures in the internet of things for smart city infrastructure: A state of art. *Transactions on Emerging Telecommunications Technologies, 34*, 11. https://doi.org/10.1002/ett.4571
69. https://media.kasperskycontenthub.com/wp-content/uploads/sites/43/2018/12/13084354/ChargePoint-Home-security-research_final.pdf
70. https://www.statista.com/topics/2430/smart-homes/#topicOverview
71. Hernandez, G., & Buentello, D. (2014). *Smart nest thermostat a smart spy in your home.*
72. Thomasian, N. M., & Adashi, E. Y. (2021). Cybersecurity in the internet of medical things. *Health Policy and Technology, 10*(3), 100549. https://doi.org/10.1016/j.hlpt.2021.100549
73. Hassija, V., Chamola, V., Bajpai, B. C., Naren, N., & Zeadally, S. (2021). Security issues in implantable medical devices: Fact or fiction? *Sustainable Cities and Society, 66*, 102552. https://doi.org/10.1016/j.scs.2020.102552
74. https://www.cisa.gov/news-events/ics-medical-advisories/icsma-17-241-01
75. Internet of things use cases for the healthcare industry. (2020). *Springer eBooks*. https://doi.org/10.1007/978-3-030-37526-3.
76. Cha, S., Baek, S., Kang, S., & Kim, S. (2018). Security evaluation framework for military IoT devices. *Security and Communication Networks, 2018*, 1–12. https://doi.org/10.1155/2018/6135845
77. https://www.computer.org/publications/tech-news/research/internet-of-military-battlefield-things-iomt-iobt
78. Yushi, L., Fei, J., & Hui, Y. (2012). *Study on application modes of military Internet of Things (MIOT).* In 2012 IEEE International conference on computer science and automation engineering (CSAE). https://doi.org/10.1109/csae.2012.6273031.

79. https://www.cisa.gov/topics/cyber-threats-and-advisories/nation-state-cyber-actors
80. Krishnapriya, S., & Singh, S. (2024). A comprehensive survey on advanced persistent threat (APT) detection techniques. *Computers, Materials & Continua/Computers, Materials & Continua (Print)*, 1–10. https://doi.org/10.32604/cmc.2024.052447
81. https://cove.army.gov.au/article/internet-military-things
82. Romaniuk, S. N., Catino, M. S., & Martin, C. A. (2023). *The handbook of homeland security*. CRC Press. https://doi.org/10.4324/9781315144511
83. Kushner, D. (2013). The real story of stuxnet. *IEEE Spectrum, 50*(3), 48–53. https://doi.org/10.1109/mspec.2013.6471059
84. Farwell, J. P., & Rohozinski, R. (2011). Stuxnet and the future of cyber war. *Survival, 53*(1), 23–40. https://doi.org/10.1080/00396338.2011.555586
85. Katulić, T., & Protrka, N. (2019). *Information security in principles and provisions of the EU data protection law*. In 2019 42nd International convention on information and communication technology, electronics and microelectronics (MIPRO) (pp. 121—1225). https://doi.org/10.23919/MIPRO.2019.8757153.
86. Bastos, D., Giubilo, F., Shackleton, M., & El-Mousa, F. (2018). *GDPR privacy implications for the Internet of Things*.
87. Umeda, S. (2018). *Japan: Basic act on cybersecurity amended*. Web Page. https://www.loc.gov/item/global-legal-monitor/2018-12-26/japan-basic-act-on-cybersecurity-amended/
88. https://www.nisc.go.jp/eng/pdf/cs-senryaku2018-shousaigaiyou-en.pdf
89. https://u.ae/en/about-the-uae/digital-uae/data/data-protection-laws
90. https://tdra.gov.ae/en/About/tdra-sectors/telecommunication/regulatory-affairs-department/regulations-and-ruling
91. https://cetome.com/panorama
92. https://www.homeaffairs.gov.au/reports-and-publications/submissions-and-discussion-papers/code-of-practice
93. https://www.homeaffairs.gov.au/reports-and-pubs/files/strengthening-australias-cyber-security-submissions/internet-of-things-alliance-australia.pdf
94. https://www.cyber.gov.au/about-us/view-all-content/reports-and-statistics/asd-cyber-threat-report-july-2022-june-2023
95. https://www.globalsign.com/en/blog/cybersecurity-improvement-act-nist-iot
96. https://galendata.com/california-sb-327-the-security-of-connected-devices/

Some Aspects of Hardware Trojan in Integrated Circuits and Its Detection and Prevention

Vilas H. Gaidhane, A. R. Abdul Rajak, Jagadish Nayak, Ashutosh Mishra, and Mrinal Goswami

1 Introduction

The demand for electronic devices is increasing along with the rapid development in the manufacturing technologies. People use technological gadgets in their daily lives for various uses such as information transfer, shopping, money transactions, taking pictures, etc. [1]. Data are handled and stored by electronic devices in various banks and business companies. Electronic equipment is unavoidable and mostly used in military defence [2], healthcare [3], aeronautics, and many more [4]. People will realize in the future that advanced VLSI devices are the sources of many devices, such as industry 4.0 [5], autonomous robots [6], automatic vehicles [7], smart cities [8], and so on.

As the technology of circuit fabrication changes continuously, the advancement in the size of the components and integration of these very small components become more complex to the design engineers. Moreover, the security of the designed system hardware still remains the concern, same as software, even though hardware security issue has long been known to the world. In past few years, various researchers have studied hardware security to some extent with a focus on circuit and chip levels which make up the majority of digital and analog systems. As everyone knows, the design and manufacturing process of the VLSI circuits are highly intricate in nature. As predicted by Moore's law, the number of components becomes double on the chip after every 18–24 months and rapidly results into high volume integration on the chips commonly called as System on Chip (SoC). These chips integrate the millions of transistors and logic gates known as ultra-large-scale

V. H. Gaidhane (✉) · A. R. A. Rajak · J. Nayak · A. Mishra
Department of Electrical and Electronics Engineering, Birla Institute of Technology and Science, Pilani-Dubai Campus, Dubai, UAE

M. Goswami
Assam Down Town University, Guwahati, Assam, India

© The Author(s), under exclusive license to Springer Nature Switzerland AG 2025
A. Mishra et al. (eds.), *Hardware Security: Challenges and Solutions*,
https://doi.org/10.1007/978-3-031-81213-2_11

integration (ULSI), very large integrated circuits (VLSI), large-scale integrated circuits, and system on chip (SoC). These complicated electronic circuits are manufactured by the different companies in various parts of the world. Nowadays, due to the advancement in the fabrication technologies, manufacturing becomes more flexible. However, all these manufactured circuits are mostly exposed to the substantial security threats. The malicious circuit design in the integrated circuits is called as Hardware Trojan (HT) and is the primary reason for IC security [9–11]. In 2007 [12], Hardware Trojan was defined by the well-known manufacturing industry such as IBM as a detrimental modification due to the malicious circuits present in the original design from the fabrication stage to packaging stage. Hardware Trojan is generally responsible for undesirable output while remaining hidden or unknown to the user. Thus, a small alteration to a chip's design has the potential to cause malfunctions, failures, or even the leakage of vital chip information to the outside world. During the early 90's, US military attempted and used the concept Hardware Trojan, implemented in a printer chip to activate in the enemy's air force [13]. Moreover, the country like Syria suffered from airstrike due to malfunctioning of the aircraft. It was due to Hardware Trojan insertion in the radar [12]. Recently, Huang et al. [14] presented various aspects and challenges of Hardware Trojan in the electronic circuits. Thus, the HT attacks have become a significant VLSI circuit security risk [11, 15–17]. These attacks deal with malevolent changes made to VLSI circuits during the design or manufacturing process at unreliable fabrication facilities or design place, involving unreliable personnel, equipment, or parts. Therefore, it is necessary to study the need for hardware security, Trojan attacks, modeling, and the countermeasures related to the VLSI circuits.

1.1 Need for Hardware Security

Over time, an integrated circuit design to market structure has experienced significant modifications. Figure 1 shows the various stages of the VLSI circuit design. The manufacturing of the IC design is mainly classified into five stages. The production of the integrated circuit is mainly divided into four stages: (i) System specifications, (ii) Functional design and verification, (iii) Logic verification and design, (iv) Circuit design and verification, and (v) Physical design and testing. Here, the Hardware Trojan may be implanted at any stage of manufacturing of integrated circuits. Table 1 also shows the details of broad VLSI circuit manufacturing domains and design abstraction levels. The security of the circuits depends on the creditability of the individual design stage. Here, the system specification and delivery to market are the most trusted levels where Hardware Trojan cannot insert in the circuits. However, logic-level or gate-level abstraction includes EDA tool, gate-level netlist, HDL, or any type of module description that are more prone to Hardware Trojan, and attackers can easily insert the HTs in the circuits. However, wafer manufacturing, photolithography, photomasking, and packaging may somewhat resist

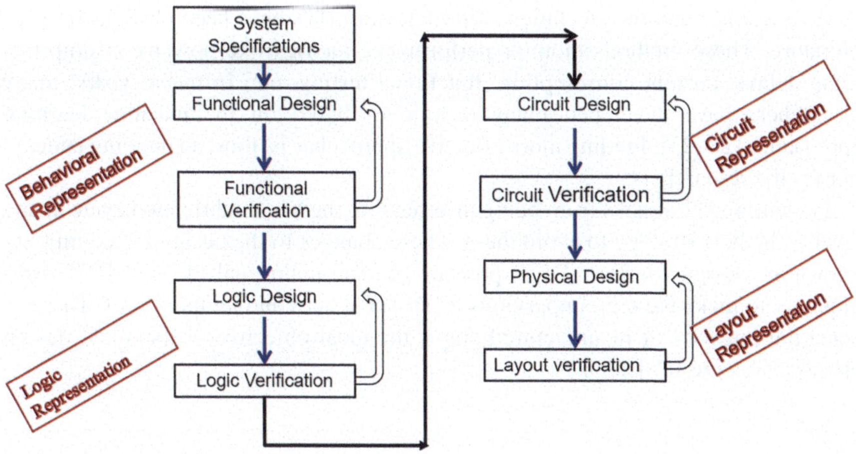

Fig. 1 Simplified VLSI circuit design flow

Table 1 Detailed VLSI circuit manufacturing domains and design levels

	Manufacturing domains		
Abstraction levels	Behavioral	Structural	Geometrical/Physical
System level	System specifications	ALU/processor	IC chip floor plan
RTL level	RTL specifications	Registers	Cells
Logic or gate level	Module description	NAND/ NOR gates	Standard cell
Circuit level	Boolean function	BJT/MOSFET	Masks

the possibility of HTs. However, in these stages, it is possible but difficult to introduce the Trojan attack in the electronic circuit.

To create a System-on-Chip, designers use testing facilities, outsourced fabrication, and Intellectual Property (IP) blocks from third-party suppliers. Because of this, there are several parties participating in the VLSI design cycle, which raises issues over the integrity of chips that are produced. This has become the prominent field of study and research for integrated circuits' hardware security. However, in this context, the classical testing approach is confronted with the few significant issues [18, 19]:

- The intricacy of contemporary IP blocks increases the challenges associated with identifying minute modifications.
- The performance of a VLSI block may not be significantly affected by the HT presence in the circuit, which might be considered as a noise in the input or output test vector.
- Since the HT cannot be located by conventional circuit tests, which are normally used to identify manufacturing flaws, they cannot ensure 100% detection accuracy coverage.

As a result, numerous techniques for detecting HTs have been developed in the literature. These methods monitor performance metrics like power consumption, route delays, current consumption, functional testing, etc. In recent years, many researchers have developed many techniques based on the machine learning approaches [20–23]. Finding more effective approaches is, thus, an ongoing endeavour for the researchers.

Preventing HT attacks from being inserted into the VLSI fabrication cycle at any level is the best strategy to avoid them. These changes to the design are commonly known as Design for Test (DFT) procedures. The main goal of these DFT techniques is to make the chip impervious to Trojan attacks and tamper-proof. To guarantee the integrity of manufactured chips, the ideal objective of the VLSI design process is secure IC design.

1.2 Current Research and Development

In order to create secure chips, many different approaches have been proposed in the past few years. A distinct degree of abstraction at the gate level is intended for each of these approaches. There is a two-stage procedure that can be imagined to guarantee the security of produced chips at the gate level. First, in what is called a vulnerability study of circuits, possible weak points in the architecture of the circuit need to be addressed. After this, the design can be modified as needed to make it resistant to tampering. Detection, prevention, and vulnerability analysis are the three primary areas of focus for security-related research. While efforts to prevent and analyze vulnerability tend to focus on the prefabrication stage, HT detection is typically performed in the post-silicon phase. A major goal of numerous studies has been the isolation of rare or difficult-to-test nets in circuit designs. Rare net detection using switching probability analysis has been the primary focus of most studies [24].

Many studies have set out to find ways to isolate circuit designs that contain unusual or difficult-to-test nets. Applying approaches based on the Sandia Controllability Observability Analysis Program (SCOAP) for "hard to test" net classification has been sparsely explored in the literature with most studies concentrating on rare net identification using switching probability analysis. Salmani and Tehranipoor [25] carried out the first study on vulnerable net detection using SCOAP-based metrics approach. In this work, the vulnerability analysis flow was performed by the authors to determine how vulnerable a circuit configuration is to HT insertion.

Based on controllability and observability assessments, Salamani [26] introduces a new Hardware Trojan detection technique in gate-level netlist. The research uses an unsupervised clustering analysis to demonstrate that in a circuit with Trojan-introduced gates. It shows that the controllability and observability properties of the Trojan gates differ significantly from those of the genuine gates, allowing for easy identification of the Trojan gates.

It is observed that the designers are able to come up with changes to the design with the help of vulnerability analysis of the gate-level circuits. These changes to the design are commonly known as design for trust procedures. There are two main categories of these methods: (i) preventative and (ii) supportive. Safeguarding the system from HT attacks is the goal of preventative techniques. The most common types of prevention approaches include logic obfuscation, split manufacturing, and layout filling. By changing the design into a functionally comparable design that is extremely difficult to reverse-engineer, obfuscation masks the original functioning of a circuit. One such mechanism is netlist obfuscation which is discussed in detail by Yu et al. in [27]. Modern techniques for hardware obfuscation are examined in this study, with a focus on those that have been developed to combat hardware Trojans and identify them.

Le et al. [28] proposed a split manufacturing that entails using both a reliable and an unreliable foundry, as the name implies. Unreliable foundries make the transistors, and the bottom metal layers make up the front end of the line layers. The designers then craft the upper metal layers in the foundry. In order to crate HT insertion, the two-stage fabrication method is effective with hiding the design parameters. The idea of split manufacturing is to fabricate partial linkages at trusted foundries, hiding crucial circuit architecture and preventing Trojan intrusion.

Typically, nonworking filler cells take up unnecessary space on the layout. Layout filling involves swapping out these inoperable cells with working standard cells. According to Xue et al. [29], a VLSI circuit consisting of these functional cells can carry out any function to be implemented for applications. This method is called Built-In Self Authentication (BISA). If you replace any of these standard cells, the signature you get from the output response analyzer will be different.

Further, Hu et al. [30] talk about post-silicon Hardware Trojan detection using gate-level information flow tracking (GLIFT). It is a design technique that stops Hardware Trojans from leaking information. All of the data flows in a circuit design can be precisely measured and managed using the GLIFT technique. Information flow tracking assigns a label to each part of data based on the level security. GLIFT is a collection of basic gates with two inputs and security labels. The GLIFT elementary gates can be used to transform any logic circuit to a gate-level netlist. Each primitive gate's logic in the netlist is traced using a constant-time mapping operation. In order to determine the signal's security level, the technique specifies rules analogous to Boolean identities. False negatives are produced when optimized circuits are subjected to GLIFT. However, an additional region overhead is produced by GLIFT.

Some scientists used HTs in the actual world and came up with a plan to protect it [31–33]. In this, authors show how two Hardware Trojans can compromise a wireless cryptographic IC with an AES core and an ultrawideband (UWB) transmitter (TX), exposing its secret key. Moreover, Fern et al. [34, 35] draw attention to the fact that existing bus implementations are vulnerable to Hardware Trojans. They offer a model for building a Trojan communication channel between the System-on-Chip (SoC) components that work with any bus topology and protocol. It is quite difficult to discover the Trojan channel because it just changes the bus signals that are already there while their behaviors are unknown. They explained the detailed advanced peripheral bus (APB) Trojan channel circuitry and built a basic system with multiple master and slave units linked via an AXI4-Lite interconnect to measure the Trojan channel overhead. Additionally, they show how a Trojan communication channel can obtain root-user data by constructing a System-on-a-chip (SoC) that runs a multiuser Linux operating system. Rudra et al. [36] explain how to implant a Trojan into a stream cipher module. However, it is observed that dealing with HTs requires a thorough and microlevel knowledge. Thus, the only way to protect HTs is to have a thorough knowledge of their traits. In recent years, different AI and machine learning-based techniques have been presented to study the HT attacks and prevention.

Liakos et al. [23] discussed about the conventional and machine learning approaches as alternative methods to detect the attacks by Trojan. Sumathi et al. [37] examined about the possible HT attacks in the PLD and ASIC life cycles in detail. They proposed cutting-edge HT detection, prevention, and diagnostic methods to the appropriate phases of the PLD and ASIC life cycles based on a comprehensive review on Hardware Trojan and their mitigation strategies. Consequently, these safety systems may be protected against HT assaults by implementing potential HT defensive solutions in their respective stages.

The review in [38] mainly focused on intrusion detection models and compares them to prevention models with the goal of reducing the impact of distributed service attacks. There has also been discussion of how intrusion detection systems are classified, various anomaly detection methods, dataset-based intrusion detection system models, data preprocessing strategies, and malware detection deep learning approaches. While addressing research difficulties, their recommended remedies, and future ambitions, a larger viewpoint has been envisioned in the paper. Moreover, pan and Mishra [39] presented an overview of machine learning-based hardware vulnerability analysis. They provided a thorough overview of machine learning-based HT vulnerability assessments. In particular, they explained about the ways in which current methods make good use of machine learning techniques to validate HT security via simulation, verification, and analysis.

Although a comprehensive literature is available, more insights are required on the benefits and drawbacks of many detection and prevention methods. Therefore, the systematic and detailed HT threats are categorized and explained in Sect. 2.

2 Classification of Hardware Trojan (HT) Potential Threats

In this section, the Hardware Trojan potential threats are classified and explained in detail. These threats are categorized in four classes: device level, system level, data level, and field level [40]. Figure 2 shows the framework of four levels of threats.

In the context of Hardware Trojans, device level threat is introduced due to the device damage. Like data, which must be physically transported, HTs are either physically implanted into the chip or altered in some way with the hardware

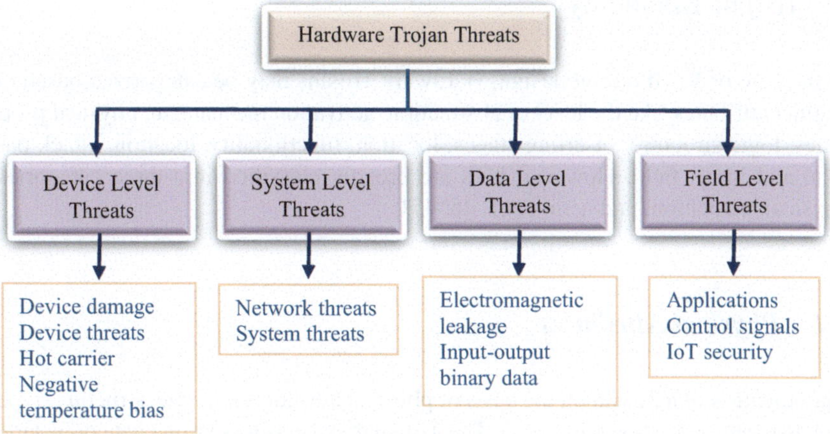

Fig. 2 Level classification of Hardware Trojan threats

settings. Thus, the Hardware Trojans inflict physical harm via the device due to faulty chip technology. Since HTs primarily aim at the bottom layer of the circuit, this kind of implantability or parasitics appropriately represents the threats characteristic at this level. Here, the negative bias temperature and instability and hot carrier may influence the reliability of the device directly and damage the circuit components of the integrated circuits. Mostly, these are the hidden threats to the devices [41].

The system-level threat is introduced due to the interference to network and devices. One way in which Software Trojans are comparable to Hardware Trojans is that they both use disguise. Moreover, it is possible to attack the electronic system by combining many HTs together. It is possible to damage the network using HT with certain components.

Attacks against data are often known as data-level threats. The leakage and subsequent manipulation of the original VLSI circuit input and output data constitute the underlying reasons for threats. Data-level threat, discussed in [42] using an elementary Hardware Trojans, occurs due to the electromagnetic leakage in the components of the circuits.

In field-level threats, Internet-of-things (IoT) security, control signals, and application layer risks are the major components responsible for the HT threat focus here. The data level's security risks are a result of interactions at the application level. It may result due to the external interference with the device components. Therefore, the good System-on-Chip device is the integral part of many applications such as smart phones, automatic cars, and humanoid robots which cannot work without the sense operation of sensors. If SoC has HT threats, it reflects on the other devices connected in the network and experience similar threats [43–45].

3 Trojan Taxonomy

In the case of VLSI circuit design, Hardware Trojans may be categorized based on distinct attributes like the level of abstraction, activation mechanism, physical properties, logic structure, inserting stages, location, functionality, location, attack pattern, and insert field. However, HTs are broadly classified into three categories: physical, activation, and action [11, 46, 47].

3.1 Physical Attributes

Classification of a Trojan on the basis of physical attributes like size, structure, type, and distribution is shown in Fig. 3. The Trojan may be as less as one gate or as large as many gates. Trojans are positioned in close proximity to one another called tight distribution. A Trojan's components may also be dispersed throughout the design. We refer to this kind of distribution as loose distribution. The trigger and payload make up the two main components of Hardware Trojan. Trigger turns on the payload by intercepting signals from the real circuit. The payload starts the unwanted effect by being patched back to the main circuit. Trojans may also be divided into two categories: parametric and functional, depending on their behavior. The functioning of the circuit is impacted by functional Trojans, which are implemented by the addition or removal of gates. Over time, parametric Trojans alter the circuit by changing device parameters such as wire thickness and the transistor doping concentration.

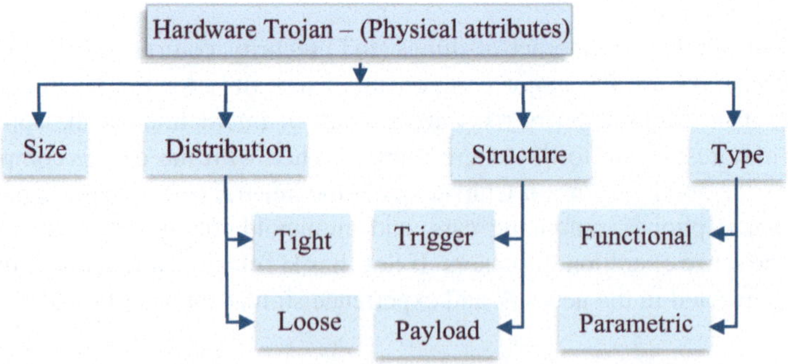

Fig. 3 HT's classification based on the physical attributes

3.2 Activation Mechanism

Activation mechanism classification falls under the category of externally initiated attacks and involves the use of devices like antenna or sensors to remotely manipulate the HTs from outside the system, as shown in Fig. 4. Because they are so obvious, always-on HTs are not very common. Keeping the HTs hidden is of utmost importance to the attacker.

An internally activated Trojan may be triggered by a particular circumstance on input lines, the internal state of a memory device, or clock signals. The activation may also be sensor-based, triggered by temperature or voltage levels. Certain Trojans are inherently persistent. They need no activation mechanism, making them very difficult to detect. Another kind of Trojans is actuated by external processes via a receiver or antenna.

3.3 Logic Structure

The third classified logic structure hardware Trojan consists of the trigger and payload [48–50]. The trigger type can be divided into digital and analog circuits. The logic structure-level Trojans may be classified according to the structural components of the Trojan, as seen in Fig. 5. Combinational and sequential circuits represent the two main categories of digitally activated Trojans. A combinational Trojan comprises a set of gates whereby the trigger component intercepts signals from various nodes within the authentic circuit to activate the payload. The signals likely to be selected are those whose logic levels fluctuate. A sequential Trojan incorporates memory components with logic gates from the authentic circuit. This complicates

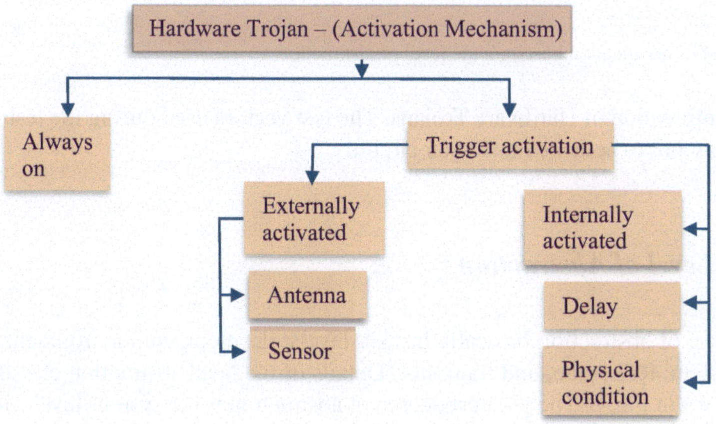

Fig. 4 HTs classification based on the physical attributes

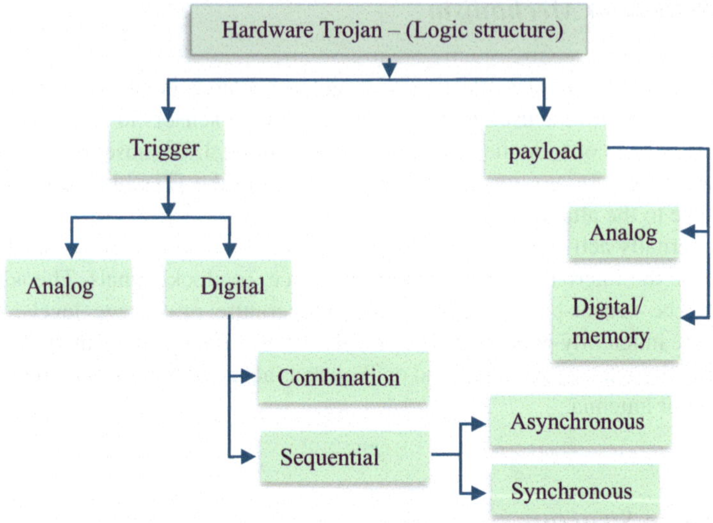

Fig. 5 HT classification based on the logic structure-level attributes

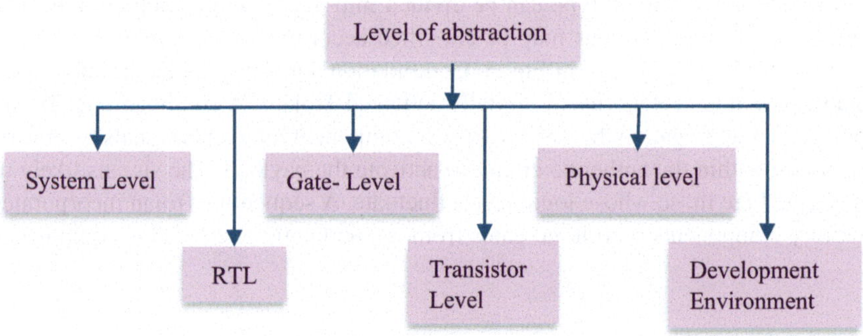

Fig. 6 HT classification based on the level of abstraction

the identification of Hardware Trojans. The test vectors used during the testing process may fail to activate the Trojan circuitry.

3.4 Level of Abstraction

The level of abstraction basically has six levels: development environment, physical, system, RTL, gate, and transistor. Details of the level abstraction classification are shown in Fig. 6. The greatest degree of abstraction is the system-level. Here, the designers will have completed all of the text descriptions they need, but since the level is not very complicated, they may add an Hardware Trojan by modifying the protocols between modules. Using the RTL (Register Transfer Level) might be risky. An attacker may implant a HT at the transistor level by changing or replacing another transistor.

4 Hardware Trojan Detection

Hardware Trojan detection methods are often formulated for a particular threat model. For more clarity, the detection methodologies can be categorized into pre-silicon and post-silicon detection. During the pre-silicon phase, the foundation of detection techniques relies on functional verification.

A significant portion of the established hardware Trojan detection techniques falls inside the post-fabrication category, concentrating on identifying hardware Trojans introduced during the foundry process. These approaches may be categorized into side channel analysis, logic testing, and intellectual property trust verification. The first methods for HT detection in this category depend on the reference models of integrated circuits. As the complexity of integrated circuit designs escalates, such models become challenging to obtain. The following subsections provide a comprehensive analysis of HT detection methodologies.

4.1 Pre-silicon Phase HT Detection

Intruders design the Trojans to stay inactive throughout the regular functioning of the integrated circuit for an extended duration. Their impact on circuitry functionality is unable to adequately assess by traditional functional verification approaches. Consequently, several sophisticated verification techniques have already been developed to identify suspicious circuitry.

Several validation approaches are developed in the literature [39, 51]. These techniques identify superfluous logic in designs by the use of a Boolean Satisfiability (SAT) solver [52]. The procedure entails the elimination of successive components in the electronic devices and the evaluation of circuits using a SAT solver. Waksman et al. [53] introduced a methodology termed FANCI for the detection of redundant logic using a measure grounded in control values. Static Boolean functional analysis is used to recognize input signals that have a minimal influence on the output. Figure 7 shows the Hardware Trojan detection details.

One subset of HT detection is performed before the final chip, known as post-silicon, while another subset is performed after the finished chip, known as pre-silicon. Thus, mainstream IC detection is categorized as either pre-silicon or post-silicon based on the IC manufacture process. Various detection techniques are aided by facilitate detection, which mostly addresses uncommon activation nets,

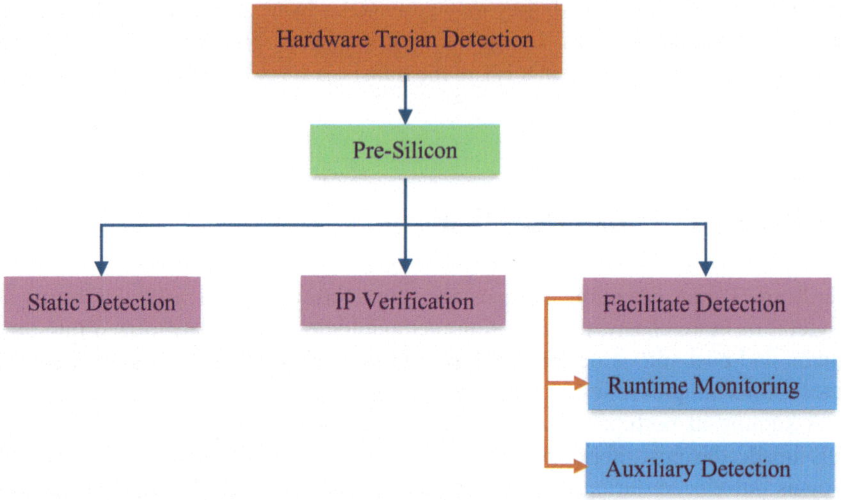

Fig. 7 Pre-silicon phase Hardware Trojan detection

processing fluctuations, and measurement error. There are two main categories: auxiliary detection and runtime monitoring.

Logic circuit examination and channel assessment may be made even more efficient using auxiliary detection. Auxiliary detection aids side-channel signal investigation via comparison of outcome enhancement based on features of HT-free IC and contaminated IC. If the system is hard for the HT to penetrate and the situation is hard for the other person to comprehend, then logical tests must be used. The term "runtime monitoring" refers to the practice of continuously keeping an eye on certain circuit components while the IC is running, which allows for the indirect detection of HT action. One benefit of this diagnosis approach is that, unlike other approaches, it provides continuous HT throughout the entire IC procedure, rather than just a single diagnosis outcome.

4.2 *Post-silicon Phase HT Detection*

It is believed that Trojans placed into a device would affect its energy consumption and route latency or electromagnetic emission (EM), so most post-fabrication HT detection methods rely on side-channel investigations. Another method for HT diagnosis is logic testing. Logic testing techniques employ suitable testing procedures with an emphasis on Trojan activity.

Depending on their size, nature, and activation mechanism, Trojans that are embedded in a design might consume electricity. No matter how much or how little activation a Trojan has, it will still suck leakage current. But, since Trojan circuitry draws so little current, it can be mistaken for noise or process changes. By taking localized current readings from several power ports, Trojan detection may be enhanced.

Post-silicon phase HT detection is classified as (i) destructive and (ii) nondestructive. Analyzing signals using side-channel techniques is comparable to the optical detection approach. It uses signals, such as light and heat, produced by the IC as it operates to determine whether the circuit includes HTs.

The goal of logic testing is to activate HT by creating a large amount of test data that matches certain, infrequent activation situations. In this method, the ICs are tested by applying the stimulus reaction and comparing the results to the predicted response based on the simulation results (Fig. 8).

By analyzing several types of side-channel signal–voltage, temperature, path delay, etc., the side-channel signal analysis compares test ICs to standard ICs. Based on the data gap, it is clear that this compared approach relies on the occurrence that certain side-channel signal parameters might indicate the additional circuits, though it fails to recognize device defects featuring HTs. This is done to determine whether the testing is done with or without HTs.

In reverse engineering (RE), the target integrated circuit (IC) is scanned layer by layer to reveal its physical structure. After delayering, the retrieved picture is compared to a reference or golden chip. In order to verify HT, this procedure may be carried out visually using an optical microscope or in conjunction with supplementary testing (such ML). Since HT can be seen at the physical level, detection rates with RE are greater. The approach has various drawbacks due to the damaging extraction of IC data. In the literature, Dong et al. [40] summarized the advantage and characteristics of pre-silicon and post-silicon Hardware Trojan detection methods, as presented in Table 2.

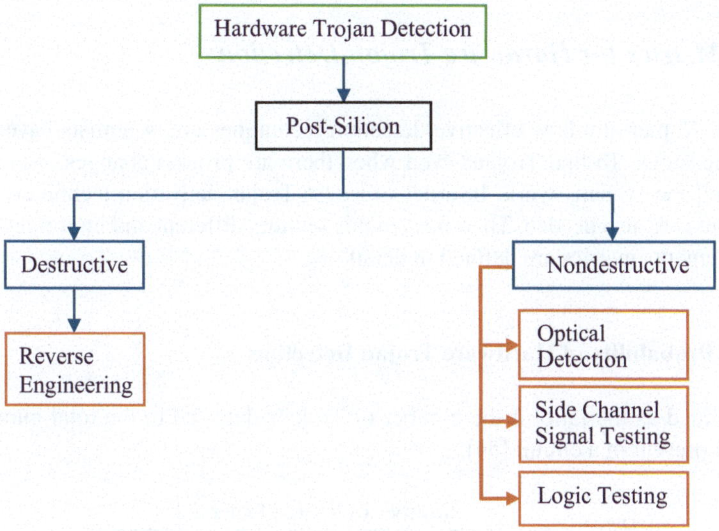

Fig. 8 Post-silicon-phase Hardware Trojan detection

Table 2 Comparison of Hardware Trojan detection techniques

Detection method	Detection approach	Technology/tools used	Advantages of technology
Pre-silicon	Static detection	Chip netlist	Comfortable for VLSI circuit, higher accuracy,
	IP verification	IP cores	Avoid the third-party attacks
	Facilitate detection	Discrete circuit components, input/output ports	Increase HT identification accuracy, higher side-channel and logic detection
Post-silicon	Optical detection	Imaging analysis optical devices	Avoid the complicated electrical and logical testing, obtain more intuitive
	Logic testing	ATP, test vectors	Higher stability and immune to noise, useful for HT chips-pin stimulus
	Side-channel signal	Precision temperature detector, oscilloscope Power analyzer, spectrum analyzer	Both detection accuracy and the number of restricted situations is high. Triggering the HTs is not necessary. The signal may be gathered if the circuit is functional. Work well with integrated chips of a smaller size. Signal can be collected.
	Reverse engineering	Electronic scanning microscope Optical scanning microscope Power analyzer Circuit analyzer Pattern analyzer	Higher detection accuracy Suitable for simple system and circuits Suitable for simple input–output data Golden chip analysis

4.3 *Metrics for Hardware Trojan Detection*

In order to measure how effective detection techniques are, scientists have developed measures. To find Trojans even when there are process changes, it evaluates how well the systems work. In order to assess Trojan activation techniques, measurements are also needed. Therefore, in this section, different and important Trojan measurements metrics are defined in detail.

4.3.1 Probability of Hardware Trojan Detection

It is defined as the ratio of the number of Trojans detected to the total number of Trojans present in a circuit [54].

$$HT_{\text{prob}} = \frac{\text{Number of Trojan Detected}}{\text{Total number of Trojan present in the circuit}} \tag{1}$$

4.3.2 Detectability of Hardware Trojan Detection

The Trojan detectability measure has been defined by Salmani et al. [55] to measure how difficult it is to detect a Trojan at the gate level. The increased capacitance caused by the Trojan gates and the number of transitions in the Trojan circuit define the measure. The capacity to identify Trojans at the level of the gate is referred to as HT detectability. It can be represented as

$$HT_{\text{Detectability}} = \frac{\dfrac{X_{\text{Trojan}}}{Y_{\text{Trojan}}}}{\dfrac{X_{\text{Trojan Free}}}{Y_{Trojan-Free}}} \times \frac{HT_{IC}}{C_{\text{Trojan Free}}} \tag{2}$$

where X_{Trojan} is the number of transitions in the Trojan circuits

Y_{Trojan} is the size of Trojan

$X_{\text{Trojan - Free}}$ is the number of transitions in the Trojan-free circuits

$Y_{\text{Trojan - Free}}$ is the size of Trojan-free circuits

HT_{IC} is the Trojan-induced capacitance

$C_{\text{Trojan - Free}}$ is Trojan-affected path with the largest capacitance in the Trojan-free circuit.

4.3.3 Hardware Trojan-to-Circuit Ratio

According to research presented in [56], the Trojan-to-circuit ratio (TCR) shows the difference over the original and Trojan-affected parameters. Assuming there is no background noise, the detection sensitivities are entirely dependent on TCR. For example, a high value of TCR will be the outcome if the Trojan-inserted circuitry introduces a significant delay compared to the Trojan-free circuit. A higher TCR value increases the potential of Trojan detection.

5 Hardware Trojan Prevention Techniques

There are two main categories of HT preventive methods: Design for trust (DfT) and Split Manufacturing for Design (SMfD).

Table 3 Comparison of Hardware Trojan prevention techniques

Detection category	Detection method	Advantages of technology	Special comments
DfT	Static detection	High precision for low-volume circuit Security significantly enhanced	Significant additional expense and efforts More cost for testing
	Layout filling	Self-detection system Reduce the space as in SoC	Suitable large VLSI circuits
SMfD	Optical detection	Imaging analysis tools Optical devices	Obtain more intuitive
	Logic testing	Automatic testing	Application for specific triggered circuits Less noise effect

5.1 *Design for Trust (DfT)*

In most cases, design for trust is accomplished by raising the layout filling, adding more circuit components, or obfuscation routines. There will be more expense for both technologies since they need more circuitry. The major drawback of the design for trust method is the significant additional expense and efforts. Nonetheless, the technique is an excellent defensive measure for some high-precision, low-volume semiconductors. With this technology, integrated circuit security may be significantly enhanced without sacrificing performance due to heavy overhead [57].

5.2 *Split Manufacturing for Design (SMfD)*

The core principle of SMfD [58] is to split the chip manufacturing process in half part. Physical isolation for HT implantation may be successfully achieved using this strategy. But there are still problems with 2D, 2.5D, and 3D production when it comes to process manufacturing. Simultaneously, SMfD is not marketable because of its high production costs and lengthy manufacturing cycle compared to DfTs. Table 3 summarizes the Hardware Trojan prevention techniques and their advantages as given in [40].

6 Challenges and Future Aspects

There are several challenges in the field of Hardware Trojan detection. Not having a universally accepted paradigm for integrated circuits verification is the most critical one. Another significant challenge is the activation of Trojans and the impact of noise and process changes on detection. Trojans come in many shapes and sizes, and inserting more than one may make detection much more difficult. Threat in the VLSI circuits may potentially insert unforeseen attacks on the system as technology progresses. In order to guarantee the safety of produced chips, it is necessary to further reinforce safeguards against Trojan insertion in the circuits.

The worldwide market for electronic devices is now dominated by VLSI design. Every year brings new technology and developments in the detection methods used for Hardware Trojan, which are becoming more accurate, smaller, and more covert. Generally, HTs are understood with the negative aspects. They are parasitic on the credible and legitimate circuits, so that they show various malicious ways to the circuits. However, the HTs are not completely harmful to the hardware. The beneficial applications for HTs develop beneficial effects on hardware security. HT-based watermarking [59] is a technology based on the features of the Hardware Trojan. Despite the novelty of the concept, its viability demonstrates the HT's adaptability in numerous contexts and applications. HT cannot be automatically thought of as negative. In the future, HTs are expected to have more and more new uses in the VLSI technology and manufacturing.

7 Conclusions

This chapter provides a concise overview of the aspects of the Hardware Trojan in VLSI circuits. The variety of approaches to HT detection and prevention lays out a framework for HT threats based on their severity and destructiveness and stresses the significance of HT security at different levels. Another standout component of the essay is the examination of HT on upcoming chips. It supersedes the prior study, which neglects the safety concerns of biochips, artificial intelligence chips, and radio frequency integrated circuits (RF ICs) in favor of digital and analog circuits. Moreover, the Hardware Trojan taxonomy and systematic classification are presented and explained in detail. Further, the detection methodologies are classified into pre-silicon and post-silicon detection phases. The advantages and characteristics of these methodologies are explained in detail. The HT measurement metrics such as probability, detectability, and HT-to-circuit ratios are mathematically explained. The HT prevention techniques like DfT and SMfD are also summarized in the chapter. Finally, challenges and future aspects are discussed to show the future of HTs in VLSI circuit manufacturing technology.

There is both a difficulty and an opportunity in the aforementioned Hardware Trojan. A world without HTs is possible with the implementation of a comprehensive detection and prevention system/framework. There are still unanswered questions and difficulties regarding IC security, despite the fact that the general IC ecosystem is developing and demand is expanding continuously. In future, Hardware Trojan can be treated positively and can be used in numerous applications.

References

1. Badawy, W., & Julien, G. A. (2012). *System-on-chip for real-time applications*. Springer.
2. Anand, S., & Sinha, S. (2019). 24X7 Lifeline chip for soldiers. *Procedia Computer Science, 165*, 573–581.
3. Joshi, S., Thaker, V., Amaravati, A., & Shojaei-Baghini, M. (2012). Low-power low-noise analog signal conditioning chip with on-chip drivers for healthcare applications. *Microelectronics Journal, 43*(11), 828–837.
4. Lv, H., Zhang, S., Han, W., Liu, Y., Liu, S., Chu, Y., & Zhang, L. (2020). Design and realization of an aviation computer micro system based on SiP. *Electronics, 9*(5), 766.
5. Morgan, J., Halton, M., Qiao, Y., & Breslin, J. G. (2021). Industry 4.0 smart reconfigurable manufacturing machines. *Journal of Manufacturing Systems, 59*, 481–506.
6. Guo, S., Pan, S., Li, X., Shi, L., Zhang, P., Guo, P., & He, Y. (2017). A system on chip-based real-time tracking system for amphibious spherical robots. *International Journal of Advanced Robotic Systems, 14*(4), 1729881417716559.
7. Bagbaba, A. C., Augusto da Silva, F., Sonza Reorda, M., Hamdioui, S., Jenihhin, M., & Sauer, C. (2022). Automated identification of application-dependent safe faults in automotive systems-on-a-chips. *Electronics, 11*(3), 319.
8. Ang, K. L. M., & Seng, J. K. P. (2021). Embedded intelligence: Platform technologies, device analytics, and smart city applications. *IEEE Internet of Things Journal, 8*(17), 13165–13182.
9. Farahmandi, F., Huang, Y., & Mishra, P. (2020). *System-on-chip security*. Springer.
10. Bhunia, S., & Tehranipoor, M. (2018). *The hardware trojan war*. Springer.
11. Tehranipoor, M., & Koushanfar, F. (2010). A survey of hardware trojan taxonomy and detection. *IEEE Design & Test of Computers, 27*(1), 10–25.
12. Mitra, S., Wong, H. S. P., & Wong, S. (2015). The Trojan-proof chip. *IEEE Spectrum, 52*, 46–51.
13. Li, X. W., Wei, Y. H., Zhang, Y., & Wang, X. H. (2017). *Research on detection technology and protection of hardware trojan*. In: Proceedings of international conference on computer science and technology (pp. 439–447).
14. Huang, Z., Wang, Q., Chen, Y., & Jiang, X. (2020). A survey on machine learning against hardware trojan attacks: Recent advances and challenges. *IEEE Access, 8*, 10796–10826.
15. Adee, S. (2008). The hunt for the kill switch. *IEEE Spectrum, 45*(5), 34–39.
16. Kumagai, J. (2000). Chip detectives. *IEEE Spectrum, 37*(11), 43–48.
17. Agrawal, D., Baktir, S., Karakoyunlu, D., Rohatgi, P., & Sunar, B. (2007). *Trojan detection using IC fingerprinting*. In Proceedings of the IEEE symposium security privacy (pp. 296–310).
18. Thakur, G., Jain, S., & Sohal, H. (2022). Current issues and emerging techniques for VLSI testing-A review. *Measurement Sensors, 24*, 100497.
19. Mirabella, N., Ricci, M., Calà, I., Lanza, R., & Grosso, M. (2021). Testing single via related defects in digital VLSI designs. *Microelectronics Reliability, 120*, 114100.
20. Gubbi, K. I., Saber Latibari, B., Srikanth, A., Sheaves, T., Beheshti-Shirazi, S. A., Sai Manoj, P. D., Rafatirad, S., Sasan, A., Homayoun, H., & Salehi, S. (2023). Hardware trojan detection using machine learning: A tutorial. *ACM Transactions on Embedded Computing Systems, 22*(3), 1–26.
21. Pan, Z., & Mishra, P. (2021). *Automated test generation for hardware trojan detection using reinforcement learning*. In: Proceedings of the 26th Asia and South Pacific design automation conference (pp. 408–413).
22. Yasaei, R., Chen, L., Yu, S. Y., & Al Faruque, M. A. (2022). Hardware trojan detection using graph neural networks. *IEEE Transactions on Computer-Aided Design of Integrated Circuits and Systems.*
23. Liakos, K. G., Georgakilas, G. K., Moustakidis, S., Sklavos, N., & Plessas, F. C. (2020). Conventional and machine learning approaches as countermeasures against hardware trojan attacks. *Microprocessors and Microsystems, 79*, 103295.

24. Zou, M., Cui, X., Shi, L., & Wu, K. (2017). Potential trigger detection for hardware trojans. *IEEE Transactions on Computer-Aided Design of Integrated Circuits and Systems, 37*(7), 1384–1395.
25. Salmani, H., & Tehranipoor, M. M. (2016). Vulnerability analysis of a circuit layout to hardware trojan insertion. *IEEE Transactions on Information Forensics and Security, 11*(6), 1214–1225.
26. Salmani, H. (2016). COTD: Reference-free hardware trojan detection and recovery based on controllability and observability in gate-level netlist. *IEEE Transactions on Information Forensics and Security, 12*(2), 338–350.
27. Yu, Q., Dofe, J., & Zhang, Z. (2017). *Exploiting hardware obfuscation methods to prevent and detect hardware trojans*. In: IEEE 60th International midwest symposium on circuits and systems (MWSCAS) (pp. 819–822).
28. Li, M., Yu, B., Lin, Y., Xu, X., Li, W., & Pan, D. Z. (2018). A practical split manufacturing framework for trojan prevention via simultaneous wire lifting and cell insertion. *IEEE Transactions on Computer-Aided Design of Integrated Circuits and Systems, 38*(9), 1585–1598.
29. Xue, M., Gu, C., Liu, W., Yu, S., & O'Neill, M. (2020). Ten years of hardware Trojans: A survey from the attacker's perspective. *IET Computers and Digital Techniques, 14*(6), 231–246.
30. Hu, W., Ardeshiricham, A., & Kastner, R. (2021). Hardware information flow tracking. *ACM Computing Surveys (CSUR), 54*(4), 1–39.
31. Swierczynski, P., Fyrbiak, M., Koppe, P., Moradi, A., & Paar, C. (2017). Interdiction in practice-Hardware Trojan against a high-security USB flash drive. *Journal of Cryptographic Engineering, 7*, 199–211.
32. Boraten, T., & Kodi, A. (2018). Mitigation of Hardware Trojan based denial-of-service attack for secure NoCs. *Journal of Parallel and Distributed Computing, 111*, 24–38.
33. Fern, N., San, I., Koc, C. K., & Cheng, K. T. (2017). Hiding Hardware Trojan communication channels in partially specified soc bus functionality. *IEEE Transactions on Computer-Aided Design of Integrated Circuits and Systems, 36*, 1435–1444.
34. Liu, Y., Jin, Y., Nosratinia, A., & Makris, Y. (2017). Silicon demonstration of Hardware Trojan design and detection in wireless cryptographic ICs. *IEEE Transactions on Very Large Scale Integration (VLSI) Systems, 25*, 1506–1519.
35. Mossa, S. F., Hasan, S. R., & Elkeelany, O. (2017). Hardware trojans in 3-D ICs due to NBTI effects and countermeasure. *Integration, the VLSI Journal, 59*, 64–74.
36. Rudra, M. R., Daniel, N. A., Nagoorkar, V., & Hoe, D. H. K (2014). *Designing stealthy trojans with sequential logic: a stream cipher case study*. In: Proceedings of the 51st ACM/EDAC/IEEE design automation conference (DAC) (pp. 1–5).
37. Sumathi, G., Srivani, L., Murthy, D. T., Madhusoodanan, K., & Murty, S. A. V. S. (2018). A review on HT attacks in PLD and ASIC designs with potential defence solutions. *IETE Technical Review, 35*, 64–77.
38. Mishra, N., & Pandya, S. (2021). Internet of things applications, security challenges, attacks, intrusion detection, and future visions: A systematic review. *IEEE Access, 9*, 59353–59377.
39. Pan, Z., & Mishra, P. (2022). A survey on hardware vulnerability analysis using machine learning. *IEEE Access, 10*, 49508–49527.
40. Dong, C., Xu, Y., Liu, X., Zhang, F., He, G., & Chen, Y. (2020). Hardware trojans in chips: Survey for detection and prevention. *Sensors, 20*(18), 5165.
41. Shiyanovskii, Y., Wolff, F., Rajendran, A., Papachristou, C., Weyer, D., & Clay, W. (2010). *Process reliability based trojans through NBTI and HCI effects*. In: Proceedings of NASA/ESA conference on adaptive hardware and systems (pp. 15–18, 215–222).
42. Xie, H., En, Y. F., & Wang, L. W. (2013). Design and detection of hardware trojan based on electromagnetic leakage. *Journal of Guangdong University of Technology, 30*, 70–74.
43. Dong, C., He, G. R., Liu, X. M., Yang, Y., & Guo, W. Z. (2019). A multi-layer hardware trojan protection framework for IoT chips. *IEEE Access, 7*, 23628–23639.
44. Khalid, F., Hasan, S. R., Zia, S., Hasan, O., Awwad, F., & Shafique, M. (2020). MacLeR: Machine learning-based runtime hardware trojan detection in resource-constrained IoT edge

devices. *IEEE Transactions on Computer-Aided Design of Integrated Circuits and Systems, 39*(11), 3748–3761.
45. Guo, S., Wang, J., Chen, Z., Li, Y., & Lu, Z. (2020). Securing iot space via hardware trojan detection. *IEEE Internet of Things Journal, 7*(11), 11115–11122.
46. Wang, X., Tehranipoor, M., & Plusquellic, J. (2008). *Detecting malicious inclusions in secure hardware: Challenges and solutions*. In Proceedings of the IEEE international workshop on hardware-oriented security and trust Anaheim (pp. 15–19).
47. Rad, R. M., Wang, X. X., Tehranipoor, M., & Plusquellic, J. (2008). *Power supply signal calibration techniques for improving detection resolution to hardware trojans*. In Proceedings of the 2008 IEEE/ACM international conference on computer-aided design (pp. 632–639).
48. Bhunia, S., Hsiao, M. S., Banga, M., & Narasimhan, S. (2014). Hardware Trojan attacks: Threat analysis and countermeasures. *Proceedings of the IEEE, 102*, 1229–1247.
49. Sengupta, A. (2017). Hardware vulnerabilities and their effects on CE devices: Design for security against trojans. *IEEE Consumer Electronics Magazine, 6*, 126–133.
50. Moein, S., Khan, S., Gulliver, T. A., Gebali, F., & El-Kharashi, M. W. (2015). *An attribute-based classification of hardware trojans*. In: Proceedings of the international conference on computer engineering & systems (ICCES) (pp. 351–356).
51. Mukherjee, R., Rajendran, S. R., & Chakraborty, R. S. (2022). A comprehensive survey of physical and logic testing techniques for hardware Trojan detection and prevention. *Journal of Cryptographic Engineering, 12*(4), 495–522.
52. Alyahya, T. N., Menai, M. E. B., & Mathkour, H. (2022). On the structure of the Boolean satisfiability problem: A survey. *ACM Computing Surveys (CSUR), 55*(3), 1–34.
53. Waksman, A., Suozzo, M., & Sethumadhavan, S. (2013). *FANCI: Identification of stealthy malicious logic using Boolean functional analysis*. In: Proceedings of the ACM SIGSAC conference computer and communications security (pp. 697–708).
54. Huang, K., & He, Y. (2019). Trigger identification using difference-amplified controllability and dynamic transition probability for hardware trojan detection. *IEEE Transactions on Information Forensics and Security, 15*, 3387–3400.
55. Salmani, H. (2017). *Hardware trojan attacks and countermeasures*. In Fundamentals of IP and SoC security: Design, verification, and debug (pp. 247–276).
56. Hu, W., Chang, C. H., Sengupta, A., Bhunia, S., Kastner, R., & Li, H. (2020). An overview of hardware security and trust: Threats, countermeasures, and design tools. *IEEE Transactions on Computer-Aided Design of Integrated Circuits and Systems, 40*(6), 1010–1038.
57. Yasin, M., Mazumdar, B., Rajendran, J., & Sinanoglu, O. (2019). Hardware security and trust: Logic locking as a design-for-trust solution. In *The IoT physical layer: Design and implementation* (pp. 353–373).
58. Patnaik, S., Ashraf, M., Sinanoglu, O., & Knechtel, J. (2019). A modern approach to IP protection and trojan prevention: Split manufacturing for 3D ICs and obfuscation of vertical interconnects. *IEEE Transactions on Emerging Topics in Computing, 9*(4), 1815–1834.
59. Shayan, M., Basu, K., & Karri, R. (2019). Hardware Trojans inspired IP watermarks. *IEEE Design & Test, 36*, 72–79.

Index

© The Editor(s) (if applicable) and The Author(s), under exclusive license to Springer Nature Switzerland AG 2025
A. Mishra et al. (eds.), *Hardware Security: Challenges and Solutions*, https://doi.org/10.1007/978-3-031-81213-2

The manufacturer's authorised representative in the EU is Springer Nature Customer Service Centre GmbH, Europaplatz 3, 69115 Heidelberg, Germany. If you have any concerns regarding our products, please contact ProductSafety@springernature.com

Printed and bound by CPI Group (UK) Ltd, Croydon, CR0 4YY

30/09/2026

02225596-0003